Making Space

LDCC Learning, Development, and Conceptual Change
Lila Gleitman, Susan Carey, Elissa Newport, and
Elizabeth Spelke, editors

Making Space

The Development of Spatial Representation and Reasoning

Nora S. Newcombe
and
Janellen Huttenlocher

A Bradford Book
The MIT Press
Cambridge, Massachusetts
London, England

This book was set in Palatino by Wellington Graphics and was printed and bound in the United States of America.

Library of Congress Cataloging-in-Publication Data

Newcombe, Nora S.
  Making space : the development of spatial representation and reasoning / Nora S. Newcombe and Janellen Huttenlocher.
    p. cm. — (Learning, development, and conceptual change)
  "A Bradford book."
  Includes bibliographical references and index.
  ISBN 0-262-14069-1 (alk. paper)
  1. Space perception in children. I. Huttenlocher, Janellen. II. Title. III. Series
BF723.S63 N49 2000
155.4'13752—dc21   99-087408

for Jeff and Peter

Contents

Series Foreword

This series in learning, development, and conceptual change will include state-of-the-art reference works, seminal book-length monographs, and texts on the development of concepts and mental structures. It will span learning in all domains of knowledge, from syntax to geometry to the social world, and will be concerned with all phases of development, from infancy through adulthood.

The series intends to engage such fundamental questions as

The nature and limits of learning and maturation. The influence of the environment, of initial structures, and of maturational changes in the nervous system on human development; learnability theory; the problem of induction; domain-specific constraints on development.

The nature of conceptual change. Conceptual organization and conceptual change in child development, in the aquisition of expertise, and in the history of science.

Lila Gleitman
Susan Carey
Elissa Newport
Elizabeth Spelke

Acknowledgments

We began this book in the summer of 1993, and it is in the summer of 1999 that we are sending the final version to The MIT Press. There are many people and institutions to thank for support and encouragement during these six years of writing.

The initial drafts of the first four chapters were written during a study leave in the year 1993–94, awarded to Nora Newcombe by Temple University. Martha Farah at the University of Pennsylvania provided a home during that year, as well as exposure to fascinating research in cognitive neuroscience and a congenial group of people with whom to have lunch. Thanks to Martha, and also to Thad Polk, Lynette Tippett, Cathy Reed, Kevin Wilson, and Dan Kimberg for a great year.

We were awarded research grants from the National Science Foundation for the years 1993 to 1996 and 1996 to 1999. Without this money, many of the pages of this book would be blank. Thanks to NSF and to the taxpayers of America for funding the investigations we conducted during these years as well as the writing of this book.

We have had many collaborators. During the period of writing this book, these people included Elisabeth Sandberg, Marina Vasilyeva, Amy Learmonth, Eunhui Lie, Sarah Johnson, Judy Wiley, and Anna Drummey. In Philadelphia, from the winter of 1997 on, Camille Rocroi coordinated the work of the Temple Infant Labs with grace and humor. In addition many other undergraduate and graduate students helped with various aspects of the studies. Thanks to everyone who worked with us to decipher various parts of the puzzle of spatial development.

Several people read drafts of the book or parts of it, and provided comments both critical and congratulatory. Kathy Hirsh-Pasek read the entire book, some chapters more than once. Bill Overton read and argued over innumerable versions of chapters 1 and 8. David Uttal provided a very careful critique of chapter 6. David also read the entire book for The MIT Press, as did Jodie Plumert and an anonymous reviewer. Thanks to all these thoughtful people for guidance and the prevention of various kinds of errors.

A book requires the assembly of many materials: references, illustrations, permissions and so on. Thanks to Sarah Johnson for help with this work, as well as to the staff of The MIT Press.

The final thanks is to explain the dedication. Jeff Lerner and Peter Huttenlocher did not help with the book in a direct professional sense. But the lifetimes we have spent with them have given us the emotional and intellectual basis for our work. Thanks to Jeff and Peter, for just about everything.

Making Space

Chapter 1
Introduction

This book concerns the development of spatial representation and reasoning. Because spatial cognition is an adaptively vital skill, knowledge of how such skills develop is central to the theoretical goal of understanding human cognitive functioning and to the practical goal of optimizing development of human abilities. Historically, scholars have hoped that cognitive development could be understood in the aggregate, in terms of certain general principles. More recently though it has become clear that separate domains of development require separate analysis. While pursuing such a strategy defers consideration of how integrated functioning arises from the various cognitive tasks humans perform, and of whether there are overarching principles for development, the hope is that more certain understanding of development in particular important domains will ultimately improve the chances of a successful general analysis. At the end of this book, we will consider to what extent current understanding of spatial development in fact informs developmental theory.

In this introductory chapter we begin by discussing at a bit greater length two preliminary matters to which we have already alluded: the importance of the spatial domain, and the strategy of domain-specific analysis as a means for understanding cognitive development. We then survey the three approaches to spatial development that have dominated most prior thinking on the topic: Piagetianism, nativism, and, to a somewhat lesser extent, Vygotskyanism. Finally, we outline the plan for this book.

Why Space?

Spatial competence is a central aspect of human adaptation. Spatial knowledge is essential to life in the world, since anything concretely existing in the world must have some spatial location—perhaps not a known one, but at least a potentially knowable one. The philosophically fundamental aspect of spatial knowledge was most famously discussed by Kant. In addition spatial knowledge is fundamental from a biological point of view. In order to survive and reproduce, all mobile beings must be able to organize their action in the spatial world. Human beings presumably evolved as hunters and foragers in an ecology in which their reproductive fitness was linked to their ability to track wild animals and return home, to find edible vegetation and rejoin a larger group, and to avoid dangers from predators and from the physical environment.

As tool use and artifact making became part of the human repertoire, reproductive advantage likely accrued from being able to imagine and construct useful implements and materials. In the current world, spatial competence is basic to daily activities such as assembling breakfast, walking to work, or fitting large objects into the trunk of a car, and also to higher-level activities such as sophisticated mathematical thinking, using information presented in graphs, diagrams, maps, and other spatial layouts (e.g., diagnostic imaging), and understanding verbal descriptions of spatial material (e.g., as when following instructions on hooking up electronic equipment).

Thus, to understand human cognitive functioning, we must understand how people code the locations of things and navigate around the world, and how they represent and mentally manipulate spatial information. To be successful, spatial coding systems and spatial representations must be based on physical principles that structure the material world. Without at least tolerably close correspondence between internal representations and the actual physical world, we would not be able to find what we need, avoid what we fear, or imagine and construct real tools.

Domain-Specific Analysis

Spatial competence is not only an important aspect of human intelligence, it is also a distinct aspect, separable from other cognitive activities at the behavioral, computational and neurological levels of analysis. At the behavioral level, both psychometric and experimental evidence have led to the identification of distinct spatial representations and thought processes. A spatial factor is one of the most consistent factors to emerge from factor analytic studies of intelligence (e.g., see Carroll 1993), and spatial reasoning seems to involve analogue processes distinct from those involved in verbal reasoning (e.g., see Shepard and Cooper 1982). At the computational level, recent work indicates that while qualitative representations alone are sufficient for certain problems, metric representations are required for successful spatial reasoning by machine (Forbus, Nielsen, and Faltings 1991). At the neurological level, spatial functioning is known to involve distinct brain areas (i.e., the hippocampus, parietal cortex, and areas of prefrontal cortex), as shown by studies involving imaging techniques, single-cell recording, and effects of brain damage. For instance, selective damage to the posterior superior parietal lobes can result in specific spatial deficit, as when a patient can "see" objects clearly (i.e., recognize them, name them, and reach for them when they are visually present) but cannot locate them (i.e., cannot reach for objects with eyes closed or after the objects have been absent for more than a few hundred milliseconds; Stark, Coslett, and Saffran 1996).[1]

Progress in understanding cognitive development requires domain-specific analysis of distinct aspects of intelligence, such as space and language, for several reasons. First, we know that certain capabilities in the adult—including spatial location, as well as language and face recognition to give other examples—are supported by distinct neurological substrates. Charting the developmental course of such neural organization is one of the goals of developmental analysis, and this goal entails domain specificity. In addition knowledge of the mature neural organization usually provides important clues and constraints for theorizing at the behavioral level, and vice versa. Second, characterization of mature competence is a prerequisite to analyzing starting points and developmental change, and such characterization must inevitably be done in domain-specific terms. It makes no sense to speak of mental rotation when thinking about language development nor to speak of the grammatical category of subject when thinking about spatial development. Third, the nature of the information that leads to developmental change differs across domains. Much has been made, for instance, in the study of language development of the lack of feedback to children regarding their syntactic errors (e.g., Pinker 1984, 1994). But this is surely not true of spatial development. Children either find a lost object they are searching for or they don't; they either get lost or they don't.

When we speak of domain-specific analysis, we invite several questions. One of these is: What exactly is a domain? There are many different definitions, associated with very different theoretical perspectives, ranging from Fodor's (1983) modularity to expertise-based areas of developed knowledge, such as ability to play chess (Wellman and Gelman 1992). Implicit in what we have written so far is a view of how to identify domains likely to be productive for developmental analysis. What we mean by a domain is a set of behaviors that meet certain

---

1. In everyday life, the patient had difficulty finding her way around familiar environments, including even her own house.

criteria: they are important for survival and reproduction, philosophically fundamental, behaviorally and computationally independent of other abilities, and dependent on neurologically specialized mechanisms in the mature brain. These criteria often converge, for good reason. Ontologically fundamental understandings are good candidates for functions that the cognitive system has evolved to perform so as to ensure survival and reproduction. Such functions may be performed by special-purpose neural circuitry in mature organisms, and may follow individual ontogenetic laws.

Note that a commitment to domain-specific analysis does not, in our view, entail a commitment to the idea that there exist innately available autonomously running sets of procedures for dealing with specific adaptive problems (cf. Fodor 1983). Dedicated neural systems in adults, as well as universally observed developmental sequences, may emerge from interaction of biologically based starting points with the environments that infants inevitably or almost inevitably encounter after birth. Such a view is quite different from hypothesizing direct specification at birth of particular representations or procedures (Elman et al. 1996; Thelen and Smith 1994). Development can lead to emergent modularity, but it can go the other way as well. Karmiloff-Smith (1992) proposed that there sometimes exists an early modularity that is punctured in the course of development, such as when perceptual codings become accessible to symbolic representation and verbal description during childhood.

There are dangers as well as virtues to a domain-specific approach to cognitive development. Most notably, investigators of a single domain risk focusing in increasing detail on that domain in isolation from all other aspects of cognitive development. To be successful, a domain-specific approach must involve cognizance of other lines of development. The aim is not only to understand a domain, such as spatial development, but also to relate discoveries about development in the domain to analyses of other domains and of cognitive development more generally. Thus a domain-specific approach to cognitive development is complementary to domain-general discussions of cognitive development, which help remind us of the importance of looking for deep connections (or contrasts) among domains (e.g., Gopnik and Meltzoff 1997; Rogoff 1990; Siegler 1996).

Three Prior Approaches to Spatial Development

There are many theories pertaining to specific aspects of spatial development, such as understanding maps. In this book we will touch on most of these. For now, we want to consider only the three theories of spatial development that have exerted dominant general influences:

Piaget's thinking on space, nativism, and Vygotsky's views on cultural transmission of spatial skills. Each of these approaches has merit, but each has been found to offer only a partial view of spatial development.

Piaget's Approach to Spatial Development
As with many other domains of cognitive development, the initial account of the origins of adult spatial competence is found in Piaget. Piaget (1951, 1952, 1954) argued that infants are born without knowledge of space, and without a conception of permanent objects which occupy and structure that space. He suggested that infants begin by treating objects as defined by their own activity. In this view an object acquires an existence and a location defined by the physical action needed to obtain or manipulate the object or by the perceptual activities needed to see or hear it. From this starting point, Piaget charted the growth of more mature concepts. For instance, he claimed that initial understanding of extent involves qualitative division of the sensory world into the categories of "reachable" (or near) and "unreachable" (or far). This initial notion was said to be centered on the self; decentering (i.e., generalization of the concept of extent to encompass relations not involving the self) was seen as a gradual process, sometimes termed the egocentric-to-allocentric shift.

Writing about spatial development after infancy, Piaget and Inhelder (1948/1967) described qualitatively different stages of thought. They argued that children begin by thinking about spatial location topologically, that is, in terms of continuities and discontinuities—children see objects simply as touching one another, as enclosed one by another, or as separated from each other. More mature spatial coding was said to appear at the age of nine or ten years, in systems called *projective* and *Euclidean space*. Piaget and Inhelder defined projective space as the coding of the order of objects along different lines of projection extended from one referent object to another, and Euclidean space as the coding of objects metrically, with reference to vertical and horizontal reference lines. While projective space seems simpler than Euclidean space in that it involves ordinal rather than metric measurement, projective and Euclidean space were not clearly developmentally sequenced in Piaget's writing (see Newcombe 1989 for further discussion).

Piaget's account of spatial development has inspired a tremendous amount of productive empirical work, literally hundreds of studies of the egocentric-to-allocentric shift, infants' search for objects, preschoolers' responses to distance problems, and elementary school children's ability to copy models or imagine another person's view of a spatial array. In general, much of this research has been critical of Piaget's account of spatial development. Studies have shown that infants begin with more equipment for spatial analysis than Piaget imagined, that preschoolers can reason about distance, and that elementary school children (and even preschoolers) can reason about spatial perspectives.

Each of these research issues is thoroughly treated in this book. In very rough preview, we conclude that Piaget indeed erred in suggesting that many spatial achievements are reached quite late in childhood. We argue that findings of younger children succeeding at versions of his tasks are not always the product of simplifying the tasks to the point where they assess abilities quite different from those of interest to Piaget (as urged in Piaget's defense by Chapman 1988). Such successes tell us something about the growing points in development, just as the failures tell us what remains to be accomplished. In addition, perhaps inadvertently, Piaget erred in implying that adults are accurate in spatial tasks; important errors and biases exist in mature spatial coding and reasoning. Most important, we conclude that Piaget's analytic tools for discussing spatial development—the distinctions he made among topological, projective, and Euclidean space—are not helpful to an analysis of development.

Despite these problems Piaget made a great contribution to our understanding of development in general and spatial development in particular by recognizing that a developing individual makes use of existing cognitive understandings (beginning from some neonatal starting point) in interpreting (and indeed selectively seeking) environmental and cultural input. Jumping from criticisms of Piaget's hypotheses about cognitive development to the rejection of the constructivist project has been a common move over the past few decades, but failures to support Piaget's very specific empirical hypotheses have narrower implications than are often realized. They do not serve to condemn an interactionist theory in general, especially when the major alternatives, notably radical nativism and simple cultural transmission theory, suffer from far worse problems. We suspect that Piaget might not be distressed to be wrong about egocentrism or to give up thinking about early spatial understanding as topological. His original and central interest was in intelligence as an adaptive characteristic, by which the actions of a child with certain initial endowments within a certain physical and cultural environment can lead to the creation of knowledge.

Nativist Approaches to Spatial Development
Out of the waning of confidence in Piagetian theory came the resurgence of nativism in thinking about cognitive development. Several investigators suggested that spatial understanding may be innately

available to infants (most recently, Spelke and Newport 1998). This conclusion has been endorsed in other contexts, for instance, in Gallistel's (1990) discussion of spatial abilities in animals and in Geary's (1995) paper on gender differences in mathematical performance.

At least three kinds of argument and evidence have been taken to support a nativist approach to spatial development. Each argument, however, suffers from empirical flaws or from errors in interpretation or emphasis, as we will discuss at various points in this book but preview here. First, it has been argued that there exists early appearing ability to perform spatial analysis, independent of visual input, as shown by the performance of a young blind child on simple encoding and inference tasks (Landau, Gleitman, and Spelke 1981; Landau, Spelke, and Gleitman 1984). However, this argument depends on study of a single blind child. Other data on spatial inference in blind children suggest very different conclusions, showing spatial development proceeding more slowly than in sighted children—although constructing spatial relations on the basis of nonvisual experience is eventually possible. Overall, the literature on blind children suggests that their spatial abilities increase over time, impressively given their impoverished input, rather than that they possess early mature ability in the absence of relevant input.

Second, it has been argued that understanding space is a modular ability, in the sense of Fodor (1983). In particular, there is said to be a "geometric module" that allows for orientation in terms of the geometry of the environment but that is impervious to information from landmarks in the environment that are noticed and would be helpful. Hermer and Spelke (1994, 1996; see also Spelke 1998) report that young children, following disorientation, rely on the relative position of long and short walls in a rectangular room to find a hidden object, to the exclusion of use of landmark cues such as the fact that one wall is painted blue. In short, the use of geometric information is said to be "encapsulated," modular, and innate. However, this argument is empirically questionable. We have found evidence that when objects truly appear permanently located (a prerequisite for treating them as landmarks), young children do use them to search after disorientation (Learmonth, Newcombe, and Huttenlocher, under review).

Third, it has been argued that biological maturation of specific areas of the brain can account for whatever aspects of spatial development are not accounted for by an innate neonatal start (Diamond 1991). But an appeal to maturation assumes a unidirectional causal path not justified by current knowledge about the role of the environment in neurological development. It is as likely that neurological changes are the product of transactions with the environment as that such changes drive and limit development. In short, the facts uncovered by proponents of nativism are more consistent with an interactionism in which environmental feedback helps to form a plastic nervous system than with a position that relegates environment to the position of a mere "trigger."

Nativists often respond to constructivists (or to environmentalists) by arguing that neonatal starting points are "fundamental" or "foundational," while environmental input, albeit required for normal development, is in some sense secondary. As Carey (1991) writes, these theorists "conjecture that ordinary, intuitive cognitive development consists only of enrichment of innate structural principles" (p. 258). Such claims are matters of emphasis or focus, however, more than statements of fact (see also Overton 1998). If two elements are needed to make a third, one needs to regard both as fundamental, even if one may be more interested in one than the other. If you are lost in the woods and need both an ignition source and dry wood to start a campfire, you would be dismayed to find yourself with only one.

Nativism suffers from an excessive focus on the strength of the origins of cognitive competence, and a relative lack of interest in environmental input and later developmental change. Nativist arguments and empirical work have, however, performed an essential service to the study of cognitive development, by focusing interest on starting points and showing that they are likely considerably more specific and powerful tools than imagined by Piaget or by the classic empiricists.

Vygotskyan Views

A third strand in thinking about cognitive development over the past few decades has been a resurgence of interest in Vygotsky's thinking. Three ideas derived from Vygotsky have been prominent in research involving spatial competence. A first theme is that of "guided participation" (Rogoff 1990), the process by which children come to understand the world better as they are guided by adults or older peers. Investigations of interactions between children and their mothers as the children engaged in copying block designs (Wertsch et al. 1980) and of interactions between children and adults engaged in imaginary errand planning (Radziszewska and Rogoff 1988, 1991) have demonstrated that such guided participation may be an important part of the development of certain spatial skills.

A second thrust of Vygotskyan theorizing has been to stress what is called the situated nature of cognition (e.g., Rogoff and Lave 1984), namely the idea that cognitive effort is uniquely adapted to the demands of particular situations and can be highly specific to those situations. Several demonstrations of situation specificity have involved spatial tasks. For instance, Gauvain and Klaue (1989) found that security guards at the New York Public Library gave much better directions than did librarians, despite equivalent lengths of time on the job and amounts of work-related travel around the building, and Gauvain and Rogoff (1986) found that children's spatial knowledge of an environment was influenced by whether their orienting instructions emphasized learning a route through it or acquiring overall knowledge of the space, including off-route features.

Third, Vygotskyan investigators have focused much interest on the uniquely human ability to deal with symbolic material such as maps or diagrams (e.g., Gauvain 1993, 1995). Thinking about spatial symbol systems naturally focuses attention on the interaction of individuals with their cultural environment, from which such representational systems are acquired. Individual experience is joined with socially guided instruction in the use of invented symbolic systems. That is, students work to understand, and teachers work to transmit, facility with navigational systems, such as studied in the navigational strategies of the Puluwat Islanders (Gladwin 1970; Hutchins 1995) or the use of various mapping conventions (e.g., those involved in rendering the globe as a flat surface). Symbolic systems allow for new knowledge to be gained without direct experience. That is, symbolic systems serve as cultural amplifiers of individual intelligence, as Bruner has long argued.[2]

Research on guided participation, situation specificity and symbolic systems has been an exciting recent area of investigation, serving as a corrective to approaches in which children develop as isolated individuals or in which they develop highly general strategies and understandings. However, focusing only on these themes gives a picture of development that sometimes overemphasizes adult instruction and cultural transmission, ignoring individuals and their own efforts to construct a sensible and coherent world. It seems likely that children interact with the physical environment as individuals as well as participate in social groups. In addition it is important to remember that children often seek out and structure their social interactions, thus giving an individual twist to an apparently social kind of transmission of information. When a child creeps up to watch a weaver at work, the

2. Nativists have often counterargued that cultural transmission depends on biological preparedness (Spelke and Newport 1998). While true, such a point hardly amounts to a denial of the tremendous advantages of *not* having to reinvent the wheel. If each child born in Oceania had to learn anew to navigate, a capacity to develop spatial inference would not allow for much actual inter-island travel, at least not without a good deal of mishap.

individual component is the choice to watch, and the social component is the availability of the guide and whatever demonstrations or advice she offers.

An overemphasis on the role of the social environment in creating and molding individual development is not an inevitable feature of Vygotskyan thinking. Clearly, there can be an important role in the theory for individual children and their sense-making efforts (Rogoff 1990). Most strikingly, Vygotskyan notions of the "zone of proximal development" are quite similar to Piagetian ideas of cognitive readiness. In either case, instruction is geared to the cognitive level of the student. Thus it seems likely that a complete theory of development in any domain must include interactions of the developing child with skilled adults and with a cultural milieu, while avoiding implications that cultural transmission is all of development or that cultural transmission imprints information on a passive organism.

New Thinking about Spatial Development

The goal of the present book is to give an account of how biological preparedness interacts with the spatial environment that infants encounter after birth to create spatial development and mature spatial competence. We begin by presenting a model of mature spatial coding as involving the coding of information with respect to two different possible frames of reference, the viewer and the external environment, as well as the hierarchical combination of information coded at various grains of resolution. We propose that infants begin life equipped with substantial spatial coding abilities, but that experience as to the consequences of using various systems as they observe and move in their environment leads to adaptive changes in the conditions under which each system is used. In addition, by 16 months at least, and perhaps earlier, children have begun to show the hierarchical combination that characterizes adult spatial coding, although the spatial categories they use differ in important ways from those used by adults.

There are other crucial changes in the first few years of life as well. To take just one example, there is a notable shift in children's spatial coding toward the end of the second year as they become capable of coding location using distal landmarks in the external environment. It is possible that this shift is linked to the attainment of a new ability to do more than code and represent spatial information but also to operate on that stored information to solve problems, that is, to do spatial reasoning. Significant developmental change in spatial coding, spatial reasoning and spatial symbol use continues through elementary school as children refine their spatial categories, learn more about how to use symbolic tools such as maps and language, and gain speed and accuracy in capacities such as mental rotation.

The approach to spatial development that we advocate is, in summary, interactionist without being Piagetian. It encompasses nativism by stressing the importance of the starting points for cognitive development in early infancy while denying radical assertions that the competencies present at the beginning are foundational and that subsequent change is no more than enrichment. It encompasses cultural transmission theory by stressing the importance of passing on tools evolved by past generations, especially in the area of graphic and linguistic representations of space, while avoiding the exclusive focus on such transmission sometimes seen in Vygotskyan writing.

Plan for This Book

In discussing spatial representation and its development, it is important to begin by describing the adult state. In chapter 2, we outline a model of spatial coding in directly experienced space, based on prior work and theory. In chapter 3, we review the traditional literature on infant spatial coding, the agenda for which was set by two questions taken from Piaget: whether or not infants begin with egocentric coding of their environment and move to an allocentric coding, and whether or not infants' search behaviors, especially the famous A-not-B error, betray a lack of spatial coding. In chapter 4, we discuss spatial coding during infancy and childhood using the formulation of spatial coding systems presented in chapter 2. We consider the early origins of coding distance in continuous space, of coding location with respect to distal external landmarks, and of hierarchical combination of information. Chapter 5 moves to consider the mental processes that operate on stored spatial information. Here, we discuss the development of abilities to take the perspectives of other viewers, to make judgments of distance, and to perform logical searches in space. Chapters 6 and 7 move away from directly experienced space to consider symbolic space. Chapter 6 deals with spatial information as encoded in models and maps, and chapter 7 considers spatial information as encoded in language. In chapter 8, we summarize the evidence and arguments, and discuss our account of spatial development in relation to various approaches to cognitive development and to other domains of development, including quantitative development, theory of mind, and language acquisition.

Chapter 2

Thinking about Space

Analyzing development in any domain is easier when one has a good idea of the endpoint toward which development is aimed. Achieving consensus on an analysis is not, however, always easy—such analysis is, after all, the primary goal of the whole field of cognitive science. In the spatial domain, several very different notions of adult spatial processing have been proposed over the years, each with very different implications for developmental analysis. On the one hand, the model of adult competence long used by many developmental psychologists (and still, we think, used by many today, often in an implicit or assumed way) is that proposed by Piaget and Inhelder (1948/1967). Piaget and Inhelder described development as culminating in a mature stage in which people have metric and Euclidean spatial representations and deal with spatial problems with ease and accuracy. Of course, if one adopts this hypothesis, the mistakes children make on spatial tasks indicate important immaturity. On the other hand, many cognitive psychologists have taken a very different view of adult spatial competence, emphasizing the distortions and inconsistencies in adult spatial processing (Byrne 1979; Hirtle and Jonides 1985; Kuipers 1982; McNamara 1991; McNamara and Diwadkar 1997; B. Tversky 1981). For instance, B. Tversky (1999, p. 39) writes that "spatial knowledge is not Euclidean like actual space, [so] attempts to model spatial knowledge have relied on topological or qualitative models." If one adopts this view, children's mistakes on spatial tasks are hardly surprising, just what one would expect given the difficulties of their elders.

We devote this chapter to defining the nature of spatial coding in mature systems. We propose a third view of adult spatial competence, different from either the perfect-competence or the distorted/ nonmetric-representation view. In this way of looking at spatial coding, adults have metric representations, somewhat as Piaget thought, but their judgments are not perfectly accurate. Instead, their spatial judgments show bias, arising from the fact that spatial coding occurs at more than one level, in a hierarchically embedded system. As well as giving a view of adult ability, this approach has implications for how to analyze development. From this perspective, some errors children make on spatial tasks indicate developmental immaturity, while others indicate the existence of a mature system showing expectable biases.

Our analysis of mature spatial coding takes advantage of an emerging consensus suggesting that spatial coding is not monolithic; it can be broken down into several systems. A fundamental distinction is that the location of objects can be coded in two basically different but coordinated ways: with respect to external landmarks or with respect to the self (e.g., Gallistel 1990; McNaughton, Chen, and Markus 1991; Sholl 1995; Woodin and Allport 1998). Further distinctions are nested within these major ones.

In this chapter we begin by discussing externally referenced spatial coding, and then examine spatial coding based on self-reference. We next outline a hierarchical model of spatial coding, which concerns how people combine information across levels and types of coding. Finally, we take up four questions that may be asked about our description of spatial coding.

Externally Referenced Spatial Coding

Coding an object's location with respect to an external frame of reference involves noting its spatial relations to other objects, usually so-called landmarks that constitute long-term stable reference systems for specific areas. That is, people's knowledge of stable landmarks provides the basis for short-term spatial coding. One can remember the location of one's keys (e.g., "on the coffee table"), or search for lost objects (e.g., "I think I dropped my keys while making the catch in left field—I was about five feet from the elm tree at that point").

Landmarks are frequently perceptually salient, familiar, and/or functionally important entities. Prototypical landmarks are buildings, especially tall and distinctive ones such as the Eiffel Tower, and fixed geographic features, such as a mountain. The only functionally vital attribute for a landmark is that it should be unlikely to move. A mailbox or a boulder may serve as useful landmarks as well.

People form long-term representations of areas, but long- term knowledge of specific spaces may vary considerably in completeness. The more one visits a place and navigates around it, the more landmarks one notices, the more one codes multiple relations among those landmarks, and the more accurate and confident one's codings of distance and direction become. In fact several studies document the evolution of spatial coding of unfamiliar areas as they are explored by subjects (Evans, Marrero, and Butler 1981; Schouela et al. 1980; the

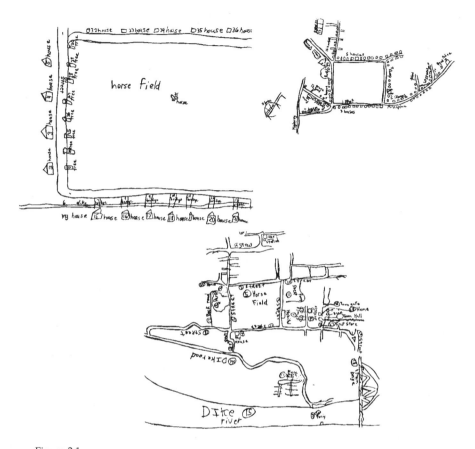

Figure 2.1
Change over six months in one 9-year-old child's sketch maps of a new environment.
From Wapner, Kaplan, and Ciottone (1981).

typical increase in detail and accuracy can be seen in figure 2.1 for a nine-year-old child). Whether detailed or sparse, the essential aspect of a reference system is that the locations of the objects in it are unlikely to change much in the short run.

Two kinds of landmarks can be distinguished: landmarks that are treated as points (e.g., an elm tree in left field) and landmarks seen as constituting a region (e.g., left field itself). Actually these types exist on a continuum, and true point landmarks exist only in the abstract, since all objects take up some space. But objects with relatively small cross sections with inaccessible insides, such as trees or flagpoles, come close to being point landmarks and are likely to be treated as such. Other landmarks, such as coffee tables or boxes, are a bit larger; objects may be located on or within such landmarks, considered as regions, or objects may be located as contiguous with them or at some distance from them, considered as points. Still larger regions, such as left field on a baseball diamond, or a room of a house, may also function as landmarks, with objects located by distance relations with respect to their borders. They can be treated as "points" as well, as when one considers something as located "in left field." In this case the categorical coding "in left field" may be taken to indicate the most central (and hence prototypical) location in the named region.

There is evidence of two dissociable systems of location coding that both utilize landmarks. *Cue learning* specifies an association between the to-be-located object and coincident landmarks (e.g., the object is under a box, on a coffee table, or somewhere in left field). *Place learning* (basically Tolman's idea of cognitive mapping) involves coding distance and direction from distal landmarks. Several kinds of evidence support the idea that cue learning and place learning are dissociable systems. First, these two kinds of learning have different maturational time courses in the rat, with cue learning appearing substantially earlier than place learning (Rudy, Stadler-Morris, and Albert 1987; Schenk 1985). Second, septohippocampal lesions disrupt place learning but not cue learning (O'Keefe et al. 1975). Third, studies using single-cell recording techniques have shown that place learning appears to be dependent on "place cells" in the hippocampus (Muller and Kubie 1987; O'Keefe and Nadel 1978; O'Keefe and Speakman 1987; Zola-Morgan, Squire, and Amaral 1986).

Cue Learning
Cue learning has sometimes been portrayed as primitive, developmentally early, and then superseded. However, it is not uncommon in everyday life even for adults. Objects frequently have habitual locations: the cookies are in the cookie jar, the earrings are in a compartment of the jewelry box, the key is under the mat, and so on. When the location is not habitual, as when one puts earrings in an empty ice cream carton in the freezer during an extended absence from home, one clearly runs the risk of forgetting the association. But assuming that the linkage is retained, the task of locating the object is achieved, and cue learning has been sufficient for the purpose.

Cue learning can also involve remembering an object as located within an area or region of a specified shape. For example, one might know that one's coat is in the closet, that one's glasses are on the table, or that one's keys were dropped in left field. Such a memory constrains search, even though it does not define a point location. For instance, a person can scan the coats in the closet for the blue one. When a location is associated with a region in space, rather than with a point landmark, people have categorical information: they know that the location is one of the large number of locations in the specified area. The most likely location, given categorical coding, is the prototypic or central location in the region.

Despite its usefulness in many situations, cue learning cannot always be used to code location. Sometimes, coincident landmarks are simply not available. If objects are located at a distance from landmarks, then metric information is needed to code location with respect to those landmarks.

Place Learning
Place learning involves specifying the distance and/or direction of a to-be-located object with respect to landmarks. Landmarks may be conceptualized either as points, as when the lost keys are thought to be five feet from the elm tree, or may be conceptualized as extended regions, as when the lost keys are thought to be certain distances from the boundaries of left field.

How much distance and direction information is required to specify the location of an object in a place learning system? Using only distance information and considering only symmetric landmarks, such as trees or towers, three non-collinear landmarks are needed to uniquely specify location, as we pointed out some time ago (Huttenlocher and Newcombe 1984). However, while some landmarks are at least roughly symmetric, many others have fronts, backs, and sides (e.g., buildings) or distinctively marked aspects (e.g., mountains with particular cliff faces or promontories). When this is the case, location can be coded in terms of a directed distance from a particular feature of that landmark (e.g., the cat is a certain distance away from the house on a line perpendicular to the front door of the house). In this case, only two landmarks are required to fix location, because angular information is encoded as well as distance information. Two landmarks are also sufficient if an object is on a line joining them and distance along this line can be coded (e.g., the dog is a certain distance from the mailbox along an imaginary line joining the mailbox and the poplar tree).

People also code location in terms of distance and direction information within a region, taking its shape and edges as a frame of reference. Such fine-grained coding may involve the implicit use of a coordinate system. So, for instance, one might consider two sides of a playground to constitute an X and a Y axis, or one might consider the home base of a baseball diamond the origin in a polar coordinate system, with the line from home base to second base establishing a 0 degree line. People code the location of a dot in a paper-and-pencil circle using polar coordinates; they code a location in a rectangle using Cartesian coordinates (Huttenlocher, Hedges, and Duncan 1991; Huttenlocher et al. 1995). Rats code metric information about the shape of a region and use it to search for food (Cheng 1986; Cheng and Gallistel 1984).

Choices of landmarks to use for location coding cannot be done idiosyncratically for different to-be-located objects if one wants to achieve integrated spatial representations. If one knows that the cat is a certain distance perpendicular to the front door of the house, and if one also knows that the dog is between the mailbox and the poplar tree, one can infer the location of the cat with respect to the dog only if one knows the relations of the mailbox and the poplar tree to the house. That is, unless different locations are encoded using the same landmarks, or landmarks with known relations to each other, it may not be possible to infer relations between the locations. A coherent spatial representation requires the use of a common system of landmarks, relatable to each other.

To speak of "coding distance" does not imply use of a formal system of measurement. Nevertheless, although specifying distance and direction in some mental measurement system does not correspond to the use of conventional systems of measurement, mental measurement "by eye" can sometimes be quite exact; people can sometimes find objects or return to locations with great accuracy. Mental measurement can also be useful even when done at rather rough grains or when associated with a good deal of uncertainty, since having even a rough idea of distance relations helps to narrow a search area to more manageable proportions.

Viewer-Referenced Spatial Coding

People must locate themselves within the external spatial world, as well as noting the relations among objects in that world. People are normally continually aware of their own location, since the information is prerequisite to acting in the world to achieve goals. Not knowing where one is can be, in both a literal and a metaphoric sense, profoundly disorienting. People who wake up not knowing where they are, for instance, while on a city-a-day tour of Europe, or people who emerge from a subway station in a strange city with no idea of where they are facing or how to begin finding a destination, may experience a feeling of panic that can only be allayed by establishing their position: by the discovery that they are in a hotel room in Florence or that they are looking north toward the Empire State Building.

Encoding the position of the moving self is an essential aspect of spatial orientation, but coding the position of objects relative to the moving self can also be the basis for spatial coding of objects (Gallistel 1990; Rieser 1989), in addition to or instead of representations using external frames of reference. As with externally referenced coding, there are two different kinds of viewer-referenced coding, one associative and somewhat limited in usefulness, and the other involving metric coding.

Response Learning
One system of viewer-referenced coding involves describing a location or a route to a location by a pattern of muscular movements that have been associated with the goal. This kind of coding is often called *response learning*, or, in the developmental literature, *sensorimotor coding*. In such learning, an acquired pattern of movements is run off in a fashion unmodified either by the organism's movement in the environment or by observation of the organism's current location with respect to an external frame of reference. For instance, a writer working at a desk may reach for a coffee cup in its usual location using such a motor habit, a person may go to the bathroom in the middle of the night following an accustomed, motorically encoded, sequence of steps, or a much-traveled route may be traversed simply by making a series of linear advances and turns.

There are important limitations of response learning. Such coding is only useful when a person is in exactly the same situation as was in effect when the motor actions that encode location were learned. Thus, when working at a new desk, the writer may be surprised and embarrassed to encounter empty space when reaching for coffee. Even if a person is in the correct spot for action to a location coded in sensorimotor terms, activation of motor sequences needs to be linked to the spatial goal of the moment. Walking a well-traveled route from work to home without thought may be shocking if one's intended destination that day was the dry cleaner.

Dead Reckoning
While response learning is limited, a more powerful viewer-centered spatial coding system exists that uses information about the self's movement. If a location is coded in terms of distance and/or direction from a person's current position, and then updated by input regarding movement (Pick and Rieser 1982), one can locate it later. Coding distances and directions of movement, using information from vestibular, kinaesthetic, and visual sources, is a powerful system of viewer-centered spatial coding, often called *dead reckoning* or *inertial navigation*. Considerable evidence exists that a variety of organisms can navigate in undifferentiated environments by retaining a memory for the extent

and direction of their movement, even with quite complicated paths (Gallistel 1990). Thus, for instance, the foraging desert ant leaves its hole and wanders around its rather uniform sandy environment in search of food; once it finds something, perhaps 100 meters from its nest after following a wandering route as long as one kilometer, it turns and proceeds quite directly back to its hole, with distance errors typically around 10% and angular errors on the order of one degree.

Research with animal models has suggested that dead reckoning may depend on compound spatial representations formed by the integration of various information sources. Exactly where these computations occur and to what regions the results are fed is an active area of investigation (e.g., McNaughton et al. 1996; McNaughton, Leonard, and Chen 1989). Cells responsive to particular head directions have been found in a variety of areas with afferent and efferent connections to the hippocampus (e.g., Mizumori 1994), but a recent study has found that hippocampectomized rats are capable of navigation by dead reckoning (Alyan and McNaughton 1999). However, dead reckoning is severely impaired in rats with lesions to posterior parietal cortex (Save and Moghaddam 1996).

Human updating of spatial coding on the basis of movement appears to take place in a relatively automatic and effortless way. For instance, Rieser (1989) has studied people who learned the locations of nine objects laid out in a circle around them. He found that they were better at indicating the location of an object from a specified facing direction when they were actually rotated the appropriate amount, as compared with situations where they only imagined rotating (see also Farrell and Robertson 1998; Presson and Montello 1994; Wang and Simons 1997). In addition Rieser, Guth, and Hill (1986) found that adjusting spatial location when actually moving is accomplished as easily for a set of several locations as for one location. When people were shown five locations in an unfamiliar room, blindfolded, led along a J-shaped path and asked to aim a pointer at one of the locations, they did as well when they had not been told ahead of time which location would be the target as when they had been told.

There may be important limits on human dead reckoning abilities, however. Studies of humans have typically found much more difficulty with path integration than seems evident in organisms such as the desert ant. For instance, blind-folded human adults are accurate in estimating location when they walk 15 meters or less, but beyond that distance they show slight but persistent underestimation effects (Fukusima, Loomis, and Da Silva 1997). Pessimism about human dead reckoning may, however, be particular to the experimental conditions used so far (for a review, see Loomis et al. 1998).

Integrating Dead Reckoning and Place Learning
A spatial coding system based on dead reckoning has both strengths and weaknesses as compared to a place learning system. The updating system in dead reckoning is apparently automatically driven by internally registered information regarding distance and direction, and dead reckoning is useful in acquiring spatial location information in conditions in which place learning is difficult or impossible, namely perceptually impoverished environments (e.g., for the blind, in the dark, or where landmarks are unavailable, as may be true at sea or in the desert). On the other hand, there will inevitably be "drift" in a dead reckoning system, so that small initial errors will concatenate and grow greater over time. That is, if one turns 40 degrees to one's left, say, but one encodes the turn as 45 degrees, then a second turn, intended to be 75 degrees, will be off not only by whatever error is associated with that turn but also by the 5 degree error carried over from the first mistake. In addition, as we have just seen, humans at least may have difficulty maintaining accurate dead reckoning over any lengthy distance.

Because of the danger of error in dead reckoning, it is generally adaptive that when the place learning and dead reckoning systems conflict, the place learning system appears to take precedence, resetting both patterns of firing in hippocampal cells and actual spatial behavior (Etienne et al. 1985; Goodridge and Taube 1995). Even when the perceived location is relatively inaccurate (e.g., when people view luminous targets in the dark), sightings of these landmarks appear to be used to guide motor activity (Philbeck, Loomis, and Beall 1997). Sailors typically use periodic "sightings" of relevant landmarks to assess whether a position as computed from the running navigational record actually corresponds with externally referenced observation, and usually minor corrections are found to be necessary (see Gallistel 1990). When sightings of external reference objects are difficult, as during long sea voyages, the consequences, prior to the invention of modern instrumentation, could be deadly (see box 2.1). Some purchase on the navigational problem can apparently be attained by the use of imaginary landmarks (see Gladwin 1970 and Hutchins 1995 for discussion of imaginary islands in Micronesian navigation, and Rieser and Frymire 1995 for an experimental demonstration).

On the other hand, when the dead reckoning system is overwhelmed by complicated and/or swift movement, and organisms experience disorientation, they may not rely on external cues. There is evidence that they rely instead on geometric cues such as the shape of a room, or else they simply re-explore the space as if it were new (Cheng 1986; Hermer and Spelke 1994 1996; Knierim, Kudrimoti, and McNaughton 1995).

Box 2.1

Launched on a mix of bravery and greed, the sea captains of the fifteenth, sixteenth, and seventeenth centuries relied on "dead reckoning" to gauge their distance east or west of home port. The captain would throw a log overboard and observe how quickly the ship receded from this temporary guidepost. He noted the crude speedometer reading in his ship's logbook, along with the direction of travel, which he took from the stars or a compass, and the length of time on a particular course, counted with a sandglass or a pocket watch. Factoring in the effects of ocean currents, fickle winds, and errors in judgment, he then determined his longitude. He routinely missed his mark, of course—searching in vain for the island where he had hoped to find fresh water, or even the continent that was his destination. Too often, the technique of dead reckoning marked him for a dead man.

From D. Sobel (1995, pp. 13–14), *Longitude: The True Story of a Lone Genius Who Solved the Greatest Scientific Problem of His Time*. New York: Walker.

Summary

As summarized in table 2.1, people may know the location of a specific object in the world in a variety of ways: by cue learning (association with landmarks), by place learning (in terms of distance and direction from point landmarks or within landmark regions), by response learning (memory for particular motor movements), or by dead reckoning (adjusting distance and direction from the self using distance and direction of one's own movement). Place learning and dead reckoning are more powerful and widely useful systems than cue learning or response learning. These spatial coding systems are normally complementary and in agreement, but when they are in conflict, the externally referenced system is usually relied on.

Combining Information across Coding Systems: Hierarchies in Spatial Coding

Spatial knowledge is inherently hierarchical. The spatial world contains a great number of features, both natural (e.g., streams, mountains, meadows and wooded areas) and constructed (e.g., buildings and streets). These can be grouped so that an area of pastures, streams, and trees may constitute, for instance, a dairy farming region, or several valleys and intervening mountains may constitute a mountain range. Similarly, a group of buildings of a certain kind constitute a neighborhood, a group of neighborhoods make up a town, and so on. In addition humans locate themselves within politically and socially defined regions such as counties, states, and countries, and these have formally defined hierarchical connections. Thus spatial environments

Table 2.1
Types of spatial coding

	Self-referenced	Externally referenced
Simple, limited	Sensorimotor learning (also called egocentric learning, response learning)	Cue learning
Complex, powerful	Dead reckoning (also called inertial navigation)	Place learning

have a hierarchical structure, in which smaller areas are related to each other and embedded in progressively larger ones. This hierarchical structure can have important effects on spatial judgments (e.g., Stevens and Coupe 1978). For instance, to use perhaps the most famous example, people are likely to believe that San Diego is west of Reno because they know that much of California is west of much of Nevada.

Given that the spatial world is inherently hierarchical, it seems natural that people's spatial codings of location at each level of that hierarchy might generally be at different degrees of spatial resolution. After all, degree of spatial resolution is one of the defining aspects of hierarchy. For instance, one may know the location of objects on a desk within a mental metric on the order of inches, whereas for a room the metric might be feet, for a neighborhood, quarters of a mile, and for a state or country, miles or even hundreds of miles.

As we have just seen, there are several ways of coding spatial information. Cue learning involves establishing an association of a to-be-located object with a visible landmark or a known region (i.e., the location is known categorically). Place learning involves using a system of estimated distance and direction with respect to point landmarks or within a region of a specifiable geometric shape. Response learning involves establishing an association of a to-be-located object with a specific motor movement or motor sequence. Dead reckoning involves coding distance and direction of one's own movement to update self-referenced location knowledge, perhaps automatically. All of these estimates may be done at various grains of approximation and may be remembered (as are all memories) with various degrees of certainty or uncertainty. And, because of the hierarchical nature of space, codings at certain levels of the hierarchy may coexist with codings at higher or lower levels. Combining all of this information, including combining information in long-term store with newly registered information, is essential to achieving a coherent spatial representation of the world.

Huttenlocher, Hedges, and Duncan (1991) presented a model in which various sources of information are combined to produce a best estimate of location. The model was originally introduced to explain a well-known phenomenon in stimulus judgment, namely a bias toward the center of categories in estimating stimulus values. The model holds that this bias arises because stimuli are coded hierarchically, at a fine-grained level (using distance and direction) and at a categorical level. When fine-grained values are inexact, information is combined across hierarchical levels in estimation. As in Bayesian procedures, which use knowledge of prior information to maximize the average accuracy of estimation, fine-grained information is weighted with category information. This process introduces bias but improves the overall accuracy of a set of judgments by reducing variability in estimates. Hence bias toward the category prototype results, not from misrepresentation of stimuli, but rather from using categories to adjust inexactly remembered information. The sources of information are themselves unbiased. This category-adjustment model has been supported in empirical studies of people's estimates of single spatial locations (Huttenlocher, Hedges, and Duncan 1991), as well as of their estimates of time and of stimulus characteristics (Huttenlocher and Hedges 1994; Huttenlocher et al., in press; Huttenlocher, Hedges, and Prohaska 1988).

Huttenlocher, Hedges, and Duncan (1991) report evidence testing the model in the context of experiments on people's coding of the location of a point in a circle. Huttenlocher et al. (1991) showed that people remember the location of such a point both at a fine-grained level (using two dimensions corresponding to a polar coordinate system), and also at a categorical level (as being in one of the quadrants of the circle formed by the horizontal and vertical axes). Bias was shown toward the category prototypes of the quadrants, located at the center of mass of each quadrant; see figure 2.2 for a schematic representation. Huttenlocher et al. found that the weighting of the prototypical category information was greater as uncertainty about the polar-coordinate location increased because there was a longer time interval between seeing the dot location and being asked to estimate it.

Kosslyn (1987; Chabris and Kosslyn 1998) suggested that the left hemisphere may be specialized for processing categorical spatial location, while the right hemisphere is specialized for processing more fine-grained spatial information. This proposal seems to provide a very natural neural substrate for the Huttenlocher et al. model. Laeng, Peters, and McCabe (1998) and Banich and Federmeier (1999) provide direct evidence that categorical coding of dots in a circle, as studied by Huttenlocher et al., is indeed more pronounced in the left than in the right hemisphere, as Kosslyn's theory would predict.

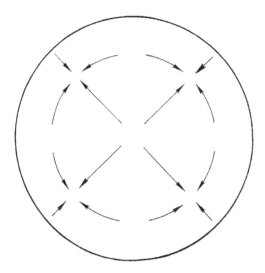

Figure 2.2
Schematic representation of bias for memory for locations in a circle. From Huttenlocher, Newcombe, and Sandberg (1994).

Categorical Coding in Natural Environments

Coding spatial location in complex natural environments is likely to involve a wider variety of different kinds of information than are relevant to coding in the circle, but the same basic principles should apply. For instance, someone who has dropped her keys on a baseball field may have both categorical knowledge of its position ("in left field") and distance and direction information with respect to land-marks (i.e., it is a certain distance from the elm tree, or a certain distance from the backstop). An estimate of the key's location will combine the distance and direction information with the categorical knowledge, with categorical knowledge weighted more heavily as the person is more uncertain about fine-grained distance and direction information. If one is quite uncertain about distance and direction memory, one may comb left field systematically beginning from the first edge one approaches; if one is more certain about distance and direction, one may concentrate initially on the small section of left field specified by that information, moving to overall inspection only if that strategy fails.

How categorical coding of location operates in natural environments does raise some issues not yet explored in the empirical literature. First, natural spaces are not always divided into regions in perceptually obvious ways. For example, while the overall shape of a playground may be defined by perceptually apparent boundaries (by surrounding streets, or by a fence), division into regions within the playground may be defined in a variety of nonperceptual ways, including functional categorization (the toddler area, the picnic area), terrain- related categorization (the sandy area, the paved area), or mental organization (as when a large grassy area is mentally divided into quadrants in a fashion similar to that used by Huttenlocher et al.'s subjects looking at small circles). Knowing what regions are used is important to being able to predict search behavior, and developmental changes in how areas are categorized may underlie some lines of spatial development.

Second, many spatial categories occurring in the natural environment do not have regular geometric shapes (e.g., the area in which one finds toddler play equipment may be asymmetrical and irregular). When regions have an irregular shape, coding location in a fine-grained way, in terms of distance from the boundaries of the region, may be more difficult than when Cartesian or polar coordinates can be applied. In addition, the size of subregions is likely to vary, rather than being equal as with the quadrants of a circle. When naturally occurring spatial regions (e.g., a house in the shape of a rectangle) have subdivisions which are not equal to each other (e.g., the rooms of the house form rectangles of varying sizes), then calibrating the scale of fine-grained coding becomes important if one wishes to combine information across various subregions.

Third, in real-world contexts, landmarks are generally present in a region, rather than the space being empty as in the paper-and-pencil circle. In fact it seems that landmarks can define spatial categories or "neighborhoods," even in the absence of physical boundaries or other perceptual cues to subdivision (e.g., the Carnegie Hall neighborhood, the playhouse area). In regions defined in this way, it might be that the defining landmark (e.g., Carnegie Hall) functions as the prototype location for the category, rather than a geometrically central location (which might be an open space or represent the location of an unimportant object). So, for a university campus defined by obvious street boundaries, a well-known landmark (e.g., a distinctive tower) might serve as a prototype location even when located near the edge of campus. On the other hand, it is certainly possible that geometrically defined centrality retains importance as a prototype, and that landmarks function as locators for the fine-grained system rather than as category prototypes. This whole issue is one which needs empirical exploration.

Fourth, the model as developed so far does not deal with situations in which people move around the world, and thus does not specifically address the use of response learning or dead reckoning in spatial estimation. Such an extension is obviously much needed.

Summary

Distance and direction information, with respect to either the self or to external landmarks, can be encoded at different levels of resolution and is known with varying degrees of certainty. In addition to fine-grained distance and direction information, people have categorical knowledge of an object's location. When people need to estimate a particular location, they combine information across types and levels, weighting each piece of information by its certainty. The result is estimates with the minimal variability possible given the information encoded, but, often, with systematic bias. This category-adjustment model of spatial location coding has so far been investigated in simple environments, but application to complex natural environments is likely possible if certain issues are explored.

Four Questions We Anticipate about This Proposal

There are at least four questions we can imagine being asked about the description of spatial coding just presented. The first question is the only crucial one, raising an issue that clearly challenges the essence of the description of spatial coding presented in this chapter. Specifically, anyone aware of the considerable body of evidence that people often exhibit systematic distortions in their spatial judgments is likely to wonder how the characterization of spatial coding in this chapter can be squared with these distortions. It is often argued that such distortions constitute clear evidence that spatial knowledge is nonmetric. We believe, however, that distortions in judgment are not only compatible with the category-adjustment model of spatial coding presented in this chapter but actually are a fundamental prediction of the model.

The other three questions are more definitional and empirical than critical, although considering them does serve to raise some important issues. One definitional question, perhaps coming from a philosopher, concerns whether the characterization of spatial coding in this chapter should be construed as taking a position in the long-standing debate about whether space is absolute or relative. Another definitional question, perhaps coming from a neuropsychologist, concerns how we distinguish "where" information, the focus in this chapter, from "what" information. A final question is an empirical one, coming most likely from a cognitive psychologist. There has been some discussion of whether people, given sufficient amounts of the right kinds of experience of an environment, can form spatial representations that are equally accessible from all possible vantage points: Do we take a position on this matter?

Why Do People Show Distortions in Their Spatial Judgments?
While we have postulated coding of metric information in both externally and viewer-referenced systems, other investigators have taken a very different view of spatial coding, portraying it as fundamentally or substantially associative, and lacking in metric information (Byrne 1979; Hirtle and Jonides 1985; Kuipers 1982; McNamara 1991; McNamara and Diwadkar 1997; B. Tversky 1981). Such models have been motivated by the observation that people make a variety of errors in spatial judgment that they seemingly should not make if they have coherent underlying spatial representations with metric properties. For instance, the distance from point A to point B is necessarily the same, in the external world, as the distance from point B to point A, an axiom referred to as the *axiom of symmetry*. Several investigators have obtained data indicating that human spatial judgments do *not* follow the axiom of symmetry, i.e., people estimate the distance from point A to point B as different from the distance from point B to point A. Asymmetries in spatial judgments have been found in a variety of environments, including urban areas, university campuses, and locations learned from a set of points presented on paper (Briggs 1973; Lee 1970; McNamara 1991; McNamara and Diwadkar 1997; Sadalla, Burroughs, and Staplin 1980). Combined with evidence of other distortions in spatial judgment, data on spatial asymmetries have led to conclusions very different from those advocated in this chapter, namely, arguments that representations of space are at least partially non-Euclidean or, in the extreme, consist merely of associative links.

We would argue, however, that findings of asymmetry in conceptual and spatial judgments do not necessarily imply a lack of correspondence between cognitive representations and characteristics of the external world. An alternative is that cognitive representations follow metric axioms and that apparent violations result from how these representations are processed. In particular, the Huttenlocher, Hedges, and Duncan model that was discussed earlier predicts asymmetry effects in spatial judgments whenever one location is closer to a prototype than the other, since locations are typically reported as being closer to category prototypes than they truly are. Thus, when the location at (or near) the prototype is fixed somewhere in space, either explicitly by an experimenter or because a person has approached the task by considering this point first, the remaining, to-be-estimated location will be placed closer to that fixed location, resulting in underestimation. On the other hand, when the location far from the category prototype is considered as fixed, the to-be-estimated location will be more accurately placed, resulting in more accurate estimation of the distance. Or, the location close to the prototype may shrink somewhat toward the prototype, resulting in overestimation. Either way, when an experimenter compares results from a condition where the location near the prototype is fixed first to those from a condition where the location far from the prototype is fixed first, an asymmetry effect will appear.

McNamara (1991) presented two possible models of spatial representation in which coding is purely or partly associative. Saying that a representation is associative does not itself, however, address how asymmetries arise. One way an associative model might predict asymmetries is in terms of how different locations function as associative retrieval cues. Such a model proposes that some landmarks are good retrieval cues that lead to the accessing of a large number of associated spatial locations, while other nonlandmarks are poorer retrieval cues that lead to the accessing of a small number of associated spatial locations (McNamara and Diwadkar 1997). Asymmetry effects arise in this model as a product of scaling effects: the same distance simply looks larger in a context in which only it and perhaps one or two other distances is included than in a context in which a larger number of distances, many of them longer, is included.

The key contrast between the Huttenlocher et al. category-adjustment model and the McNamara and Diwadkar contextual-scaling model is that the category-adjustment model holds that asymmetries arise in the process of fixing in space points which are differently located with respect to spatial categories, whereas the contextual-scaling model holds that asymmetries arise from differences in the size of the context evoked in working memory by landmarks as opposed to nonlandmarks. In the contextual-scaling model asymmetries do not depend on the spatial location of points, but on their association with other locations. There are two situations in which the models therefore make different predictions.

First, consider predictions about asymmetry effects when neither of two points is at a prototype. According to the category-adjustment model, when one point is further away from the prototype than the other, asymmetries should appear: when the closer point is fixed, the further point will shrink more toward the prototype than will the closer point when the further point is fixed. However, the category-adjustment model predicts no asymmetry effect when two points are equidistant from a prototype. In that case shrinkage of point B toward the prototype when point A is fixed should be equivalent to shrinkage of point A toward the prototype when point B is fixed. By contrast, the contextual-scaling model predicts no asymmetry effect for any pair of points when neither point is a landmark, because, on average, neither point should serve as a retrieval cue for a larger number of points than the other.

Second, consider predictions about the size of asymmetry effects as a function of distance of a point from a prototype. As long as all points are in the same category, the category-adjustment model predicts that larger asymmetries will be seen with increasing distance. By contrast, the contextual-scaling model does not predict such an effect, because there is no reason to think that the number of referents evoked by a nonlandmark would vary as a function of its position.

Newcombe et al. (1999) have shown that the under- and overestimation effects and the asymmetry effects predicted by the Huttenlocher, Hedges, and Duncan model can be observed. In particular, asymmetry effects appear even when two nonlandmarks are being probed, and asymmetry effects vary depending on the distance of a nonprototype from the prototype. Neither of these effects is easy to account for within an contextual-scaling model. (While McNamara and Diwadkar did not find asymmetry for longer-distance pairs, the likely reason for this fact is that some of the pairs involved between-category estimation rather than within-category estimation—the farther point was just too far away from the prototype to be seen as belonging at all.)

In summary, spatial coding need not be presumed to be purely associative because people exhibit asymmetry effects in spatial judgments. Asymmetries can be accounted for by a model in which coding is metric at each level but in which information is combined across levels. In fact, asymmetries are a fundamental prediction of the category-adjustment model, not an embarrassment to it.

Is Space Really Relative?
Considering the question of how location is encoded immediately raises issues concerned with defining the fundamental nature of space. Philosophical discussion has long revolved around the question of whether space is absolute or relative in nature (see Jammer 1954 for a history of this debate, and Liben 1981 and O'Keefe and Nadel 1978 for more modern accounts). In an absolute conception of space, a framework exists independent of any objects occupying it. In this view it is meaningful to use the term "empty space." Objects are located in terms of their particular places in the framework. A contrasting view holds that the concept of spatial location is inherently a relational one: entities have a location in virtue of their relations to other objects, which themselves have location in virtue of their spatial relations. No such thing as empty space is conceivable.

Our analysis of spatial coding seems inherently relational; we discuss the location of an object as encoded with respect to some referent. Whatever the nature of that referent (e.g., another object, an abstract conception of space such as a set of Cartesian coordinates), we further describe the referent as itself needing to be located, with respect to another referent or frame of reference. There are actually two potential questions concerning this view. First, logically, defining location in this way would seem to involve locating referents in terms of relations to yet other referents in an infinite regress. Second, do we really mean to adopt a relational, and reject an absolute, conception of space?

The infinite-regress problem, while logically sharp, seems to us functionally myopic. From a functional point of view, the fact that certain referents, such as geographic features or even aspects of the built environment, are apt to change very slowly and/or infrequently, if at all, renders this objection less compelling. People live in environments in which they rarely, if ever, need to locate certain features of their environment. Rather, these features are fixed and perceptually available from known vantage points. Thus these features can serve as reference points.

At the same time, even though the location of the proximal landmarks used in everyday location and navigation tasks may generally be treated as unproblematic, for certain purposes the location of landmarks such as these may also need to be fixed in terms of more extended frames of reference. That is, buildings in a city or mountains in a mountain range may be related to each other, in some cases by inferential processes in order to allow, for example, for the planning of lengthy trips or the delineation of property disputes. Whole cities or mountain ranges may in turn be related to each other through their position on the earth's surface, as when we construct a global map. Thus a landmark at one scale of resolution may be treated as a point-to-be-located at another scale of resolution because of the hierarchical nature of frames of reference. Such hierarchies probably become more extended and complex as human travel becomes more efficient, and greater areas of the world need to be related to each other. However, ultimately, a certain level in the hierarchy is reached at which, again from a functional point of view, the regress stops because the referents do not need to be located. Even an astrophysicist does not attempt to locate the universe with respect to other (hypothetical) universes.

With respect to the question of whether our approach to spatial coding entails rejection of an absolute conception of the nature of space, we think that the notion of completely empty space is an incoherent one. The word "completely" gets the stress in this rejection. An important aspect of mature spatial location coding is that although the relation of particular objects to each other or to the self may not always be known (e.g., I may not know where my house is in relation to a baseball field, or where I am currently in relation to the doctor's office I'm seeking), adults apparently know that such questions always have

Box 2.2

When Robert Gray sailed *Columbia* into the estuary of the river he named for his ship and fixed its latitude and longitude, mankind knew for the first time how far the continent extended. Knowing the exact location of the mouth of the Columbia represented a great triumph of eighteenth-century science and exploration. . . . What remained to be discovered on earth was the interior of Africa, Australia, the Arctic, and Antarctic, and the western two-thirds of North America. The latter was most important to Europeans and Americans.

It was known to be vast, some two thousand miles from the Mississippi River to the mouth of the Columbia. It was known to contain a wealth of furs. It was presumed to contain immense quantities of coal, salt, iron, gold, and silver. It was assumed that the soil and rainfall conditions were similar to those in Kentucky, Ohio, and Tennessee—which is to say, ideal for agriculture.

But what was not known, or what was assumed but was badly wrong, was more important than what was known. Donald Jackson, the great Lewis and Clark scholar, points out that, although Jefferson had the most extensive library in the world on the geography, cartography, natural history, and ethnology of that awesome *terra incognita* west of the Mississippi, when he took the Oath of Office in 1801 he believed these things: "That the Blue Ridge Mountains of Virginia might be the highest on the continent; that the mammoth, the giant ground sloth, and other prehistoric creatures would be found along the upper Missouri; that a mountain of pure salt a mile long lay somewhere on the Great Plains; that volcanoes might still be erupting in the Badlands of the upper Missouri . . . Most important, he believed there might be a water connection, linked by a low portage across the mountains, that would lead to the Pacific."

From Stephen E. Ambrose (1996, pp. 54–55), *Undaunted Courage: Meriwether Lewis, Thomas Jefferson, and the Opening of the American West.* New York: Touchstone.

some answer. Because the physical world is continuous, every point in it can ultimately be spatially related to every other point. Even when specific areas or sections are unknown, they must be considered as potentially or actually occupied, as in the maps of early explorers which contained *terra incognita* markings. Such notations serve as placeholders to indicate the relations between known areas across areas that are not known.

There is, however, an enormous difference between an area of empty space conceived of as bounded by known areas, and "empty" only because unexplored (see box 2.2 for a concrete example of such thinking in American history), and the notion of space as existing independently of any objects or boundaries at all. Attempts to imagine such a completely empty space probably covertly import the self as a hidden observer (and thus a relationally defining entity) or else import an idea of boundedness (and thus another relationally defining entity, namely, the edges of the space). Thus these imaginings are not absolute space at all.

The "Where" System and the "What" System
All objects exist in space, but this fact has two separable consequences. The first inevitable fact about physical existence is that all objects have a specifiable size and shape. Size and shape are attributes belonging to an object, basic to object identification and to interacting with that object (e.g., lifting it). We recognize the spatula or the frying pan by their shapes, and interact with them according to this recognition. The second inevitable fact about physical existence (and the one on which we focus in this book) is that all space-occupying shapes must be spatially situated in some way with respect to each other. The spatula may be on the stove beside the frying pan, or in the drawer, or in the dishwasher, and so on. Our interactions with objects require, not only that we recognize them, but also that we can find them.

One of the most intriguing advances in understanding visuospatial functioning is the discovery that there are separate systems in the primate brain corresponding to this distinction, that is, separate systems for answering the questions of what an object is and where an object is (see evidence reviewed by Ungerleider and Mishkin 1982). The "what" system involves the inferior temporal cortex; damage to this area impairs an animal's performance to make discriminations based on the appearance of objects. The "where" system involves the posterior parietal cortex; damage to this area impairs an animal's ability to discriminate on the basis of place. (Figure 2.3 shows a schematic representation of these systems.) Similar patterns of dissociation have been found in human brain damage (Bauer and Rubens 1985; Farah et al. 1988; Levine, Warach, and Farah 1985; Ratcliff 1982; Stark, Coslett, and Saffran 1996) and in studies of brain activation in normal adults (Kohler et al. 1995; Moscovitch et al. 1995).

The what/where distinction is not yet a cut-and-dried one, however. First, the what system can properly be said to encode certain kinds of where information, in that coding the shape of an object often entails coding the relative location of object parts. This kind of locational information is often thought to be encoded in an object-based (or intrinsic) frame of reference (Baylis and Driver 1993), whereas the where information for separable objects, as we have just seen, is encoded in some extrinsic frame of reference, whether observer or landmark based. However, making this distinction does not solve the basic problem of defining just where one system stops and the other starts. For instance, it is difficult to say when, in assembling various objects and interconnecting them to make an integrated "superobject" (e.g., making a new machine from assorted parts), we pass from the where to the what system. Second, it is difficult to say a priori which system is responsible for certain ambiguous spatial tasks, such as coding and

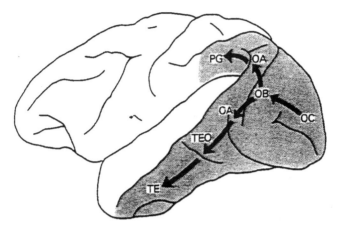

Figure 2.3
Lateral view of the left hemisphere of a rhesus monkey. The shaded area defines the cortical visual tissue in the occipital, temporal and parietal lobes. Arrows schematize two cortical visual pathways, each beginning in primary visual cortex (area OC), diverging within prestriate cortex (areas OB and OA), and then coursing either ventrally into the inferior temporal cortex (areas TEO and TE) or dorsally into the inferior parietal cortex (area PG). Figure and caption from M. Mishkin, L. G. Ungerleider, and K. A. Macko (1983), Object vision and spatial vision: Two cortical pathways, *Trends in Neurosciences*, 6: 414–17.

contrasting the size of similarly shaped objects. A case could be made for the where system: size judgments require metric information, as does coding distances between an object and an observer or a landmark. However, size also is a characteristic of a single object, and sometimes is helpful in defining what an object is (e.g., the difference between a toy spoon and a real one), an argument favoring the role of the what system. Recent evidence suggests that the information necessary for judging object size is present in both the dorsal and the ventral pathways (Dobbins et al. 1998). Dobbins et al. suggest that the existence of a common set of distance information could be the basis for the cross-referencing of computations in the two systems, something which is clearly required in many tasks. For instance, cortical areas related to both systems are activated during mental rotation (Carpenter et al. 1999).

It is also important to remember that the where system is probably not a single entity, as is (wrongly) suggested by figure 2.3. As we have seen, there are at least four major ways in which to specify the where information. The neural substrates of each system are still being worked on, but they surely involve areas in addition to the parietal cortex shown in the figure, notably the hippocampus and its allied structures.

The focus of this book is on the location of objects in the world, not on object identification, that is, on the where system not the what system, for pragmatic reasons. We acted on the not-biting-off-more-than-you-can-chew principle, not on the unimportance of the what system. Ultimately, however, a complete theory of spatial development will need to better demarcate this distinction, and deal with the where aspects of the what system.

Can Spatial Information Be Coded in an Orientation-Independent Fashion?
Some time ago, Siegel and White (1975) proposed that people initially get to know a space by noting prominent landmarks, then learn to follow routes through that space that link landmarks, and eventually acquire "survey knowledge" of the environment. Survey knowledge was conceptualized as an integrated representation from which people could read off spatial relations between points not directly experienced. Whether or not this is the case, at least in a strong sense, has come increasingly into question (although not yet into definite disrepute). There is some evidence that spatial knowledge has a good amount of orientation specificity, with novel relations computable only by mental-rotational or inferential processes that take time and have, inevitably, some proneness to error.

An initial proposal on the question of orientation-independent coding was that spatial memories formed from direct spatial experience were likely to be orientation independent, while spatial information acquired from maps was likely to be orientation specific—somewhat ironically, given the metaphor of the "survey" used by Siegel and White (1975). Several studies (e.g., Evans and Pezdek 1980; Presson 1987; Presson and Hazelrigg 1984; Sholl 1987; Thorndyke and Hayes-Roth 1982) showed that when people gained information about an environment by walking in it, they were often able to point to landmarks with equal efficiency from a variety of viewpoints. After study of a map, by contrast, people performed with much greater accuracy in tasks allowing them to remain aligned with the map in the orientation in which they learned it than in tasks that required them to imagine other headings. That is, for instance, if one has learned about the space from a conventional map with north at the top, it is relatively difficult to answer questions about directions when asked to imagine facing east, south, or west (Sholl 1987).

A distinction between symbolically acquired and environmentally learned spatial information has been supported by certain neuropsychological case studies (Schacter and Nadel 1991). Dissociations have been reported between symbolically and directly acquired spatial memories (Landis et al. 1986; van der Linden and Seron 1987). For instance, patients who were unable to orient themselves in familiar environments could nevertheless draw and utilize maps of those same environments and could remember where to go using verbal directions, for example, finding a hospital room by remembering that "the door of my room is the first one on the left after the fire escape" (Landis et al. 1986). Such data suggest that knowing an environment by symbolic means is quite different from knowing it by and during navigation.

More recent research suggests, however, that people form orientation-independent memories from direct spatial experience only when they view a large (at least room-sized) space from a more-or-less horizontal viewing angle, and are subsequently tested within the space (Sholl and Nolan 1997). These conditions are highly specific and may have led to the difficulty some researchers report in finding orientation-independent spatial memory at all (Roskos-Ewoldsen et al. 1998). Difficulty in finding orientation-independent spatial memory has led to the proposal of an alternative view of spatial memory, which holds that spatial memories are always orientation specific. Such a position is similar to the approach taken to object representation by Tarr (1995; Tarr and Pinker 1989). In this way of thinking, apparent freedom from the effects of orientation occurs only when a person has acquired a sufficient number of multiple orientation-specific views that new views can be rapidly and accurately calculated from mental rotation from a nearby memorized vantage point (Diwadkar and McNamara 1997; Roskos-Ewoldsen et al. 1998; Shelton and McNamara 1997; Sholl and Friedman 1997). Acquisition of multiple mental views may be a likely consequence of direct physical experience with an environment through which one walks along different paths and in different directions.

There is also controversy concerning the existence of multiple mental views when people form imagined scenes from verbal input. Franklin, Tversky, and Coon (1992) reported that when they asked people to evaluate spatial statements from more than one viewpoint, the participants seemed to form a single overview of the scene in which they took a neutral perspective. Maki and Marek (1997) have, however, found that people in almost identical situations seemed to take multiple egocentric perspectives. There is currently no explanation for these opposing results.

The multiple-views approach is a relatively recent one, and needs further testing. There may be ways of integrating the multiple-view theory with claims of orientation-independent coding. For the case of representations of familiar objects, Burgund and Marsolek (1998) have recently found that viewpoint-dependent priming is found in the right hemisphere and viewpoint-independent priming is found in the left hemisphere. It is certainly possible that similar findings may be obtained for multiple-object scenes. Or, it may be that the goals of the learner influence whether or not orientation-independent coding is found. Taylor, Naylor, and Chechile (1999) found, in a situation involving route versus survey learning, that representations were sometimes bound to the learned perspective and sometimes showed perspective flexibility, depending on whether people had been asked to concentrate on learning a route or learning the overall layout.

An important question is whether multiple views of a scene lead to orientation-free performance even though single views do not. Sholl and Friedman (1997) suggested that multiple views can have this effect, although Shelton and McNamara (under review) report evidence that multiple views lead simply to multiple representations. Clearly, we need to know much more about whether acquisition of a sufficiently large number of view-specific "snapshots" would eventually lead to an integrated knowledge system from which novel viewpoints could be accessed directly rather than computed, perhaps much as Siegel and White originally thought. Distinguishing the idea that people eventually form integrated representations with equipotentiality of access from the multiple-views hypothesis may be difficult, however. After all, the point of having a large number of multiple views, functionally, is presumably that novel headings could be computed with relative ease and rapidity, in a fashion difficult to distinguish from direct access to an integrated representation.

Conclusion and Preview

In this chapter we have outlined our conception of mature spatial coding. The mature system appears to comprise four possible systems for coding spatial location, two of them externally referenced (i.e., cue learning and place learning), and two of them viewer centered (i.e., response learning and dead reckoning). We have argued that it is adaptive to combine information across systems of coding as well as across hierarchical levels of spatial resolution. Such combination processes may introduce bias, but bias does not necessarily indicate fundamental distortion in underlying representations.

This analysis of spatial coding allows us to understand spatial development as growth toward such a mature system. Biases or errors that are inevitable given the system are seen as signs of maturity, whereas biases or errors different in kind from those made by adults are seen as indications of the state of the emerging competence. An advantage of this conceptualization is that it suggests specific questions about the development of spatial coding. Clearly, one wants to know whether all four fundamental spatial coding systems are evident early in life as starting points for further development or whether one or more emerge during development (and if so, when and how). In addition one wants to know how the developing child manages situations in which two or more of the coding systems are in conflict, and whether the developing child combines information hierarchically.

In the next chapter we begin to examine spatial coding capacities in infants. Unfortunately, the questions asked in most of the existing literature on the development of spatial representation in infancy are not quite the questions suggested by the analysis developed in the present chapter. Past research has mostly centered on the hypothesis that infant spatial coding begins as pure response learning, or egocentrism, making a qualitative transition to "allocentric" learning (i.e., everything else) at some point in the first year, and on the intertwined issue of whether infants conceptualize objects as having permanent locations at all. Much less is known about dead reckoning, place learning, and hierarchical coding in infancy, and there has been little attention to the general question of how developing organisms learn to weight competing sources of spatial information. In the next chapter we begin our examination of spatial development during infancy by evaluating the existing literature on the egocentric-to-allocentric hypothesis and the development of object permanence as assessed especially in the A-not-B error. We defer to chapter 4 discussion of research on early spatial development specifically inspired by the analysis of spatial coding in this chapter.

Chapter 3
Two Hypotheses about Infant Location Coding

The essential task of a theory of cognitive development in any particular domain is to delineate the starting points for development, in the capacities of infants, and to describe and explain changes from that beginning. As we said at the end of chapter 2, we face a problem in reviewing the theory and evidence regarding infant spatial capacities, in that most investigators have asked questions about early spatial development inspired by Piaget's characterization of spatial understanding in infants and adults, rather than by the analysis of mature spatial competence we developed in chapter 2. A large number of studies have examined the hypothesis that there are qualitative shifts in spatial coding in the first year of life, from an initial state of sensorimotor knowledge (*egocentric coding*) to a coding system that relates objectively to the environment (*allocentric coding*). An even larger number of studies have focused on the development of the object concept, and, in particular, on explaining an odd error sometimes occurring in infant search, the A-not-B error.

Our approach to this state of affairs is simple. In the present chapter we bow to the state of the existing literature, discussing what has been learned about spatial development by investigating the egocentric-to-allocentric shift and the A-not-B error. The A-not-B error is an especially complicated tale, since there are several important determinants of infant search for objects in this situation, with spatial coding being but one of them. We save for chapter 4 a discussion of the evidence on questions generated by a different approach to the nature of spatial coding.

The Egocentric-to-Allocentric Shift Hypothesis

Research over the past decades has generally been taken to support Piaget's claim that infants initially encode location in sensorimotor terms and subsequently go through a qualitative shift to allocentric coding as a result of their physical interactions with the world. The shift has specifically been thought to be propelled by the development of locomotion in the form of crawling (e.g., Acredolo 1990; Bertenthal and Campos 1990; Thelen and Smith 1994). While the recent focus on locomotory experience is somewhat different from Piaget's approach, since Piaget put more emphasis on grasping and the coordination of different sensorimotor schemata than on crawling, the basic approach to infant development is very similar.

There are, however, other ways to read the evidence (see Bertenthal 1996; Haith and Benson 1998; Millar 1994). Prelocomotory infants appear to have greater coding capacities than the egocentric-to-allocentric proposal acknowledges. Developmental change may consist mostly in changes in the likelihood of use of alternative types of spatial coding, caused by discoveries made by the infant about cue validities when different codings are in conflict rather than in the invention of new forms of coding. Further, locomotion, while one powerful source of developmental change in weightings, does not seem to be the only source. To understand these issues, we need to begin by explaining the basic paradigm used to study the egocentric-to-allocentric shift.

Initial Studies

Acredolo (1978) and Bremner and Bryant (1977) began the systematic study of spatial coding in infancy with somewhat similar experiments. In Acredolo's work, infants were seated in a square featureless room with windows on the left and right (see figure 3.1). A centrally located buzzer sounded, letting the infants know that they could see a person making engaging faces at them, if they turned their heads toward one of two windows. There was one window on the left wall and one on the right wall of the room; infants saw events only on the left (or only on the right) side. Following trials of this kind, the infant was moved along a semicircular path to the opposite side of the room. The crucial data concerned where the infants then looked when the buzzer sounded: to the same side of their body as previously led to the interesting event (thus ignoring the fact of movement) or to the other side (thus taking movement into account).

Bremner and Bryant (1977) used a slightly different procedure. They gave 9-month-olds experience finding an object hidden under one of two identical cloths. The cloths were placed side by side on a table, one to the left of the infant's midline and the other to the right. They then moved the infants to the opposite side of the table. The crucial data concerned whether infants searched under the cloth on the same side of their body as where they had seen the object hidden (a self-referenced coding not taking account of movement) or on the opposite side (now the correct location).

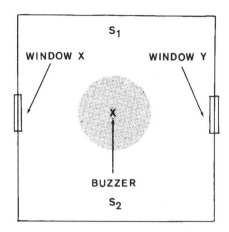

Figure 3.1
Experimental setup devised by Acredolo (1978). Infant sits at S_1 and experiences an interesting event at window X after buzzer sounds. Child is moved to S_2, and observed for direction of gaze after the buzzer sounds.

Piaget's idea of sensorimotor coding clearly predicts that infants in both kinds of experiments would search incorrectly, on the side defined by their previously correct body movements. This is the egocentric choice. Correct responding, however, is probably misleadingly termed allocentric responding, since the use of a single term might be taken to imply the existence of only one allocentric coding system. In fact a correct choice in this situation can be based on more than one kind of spatial coding. Infants can succeed by coding the distance and direction of their movement (i.e., by dead reckoning). Such coding is the only option available when infants are studied in environments whose features are carefully placed so as not to provide cues, that is, with objects placed centrally, or with duplicate objects placed symmetrically along the left-right axis. Infants can also succeed in these studies by coding location using external landmarks, when informative landmarks are available. As we discussed in chapter 2, it is worthwhile distinguishing between coincident landmarks, which allow for cue learning, and distal landmarks, which require place learning. The egocentric-to-allocentric literature has concentrated on studying coincident landmarks (i.e., cue learning). So, in summary, when infants in these studies make a nonegocentric choice, they may be using dead reckoning, cue learning, or place learning, depending on the landmarks available. When all relevant landmarks are removed or made uninformative, they must be using dead reckoning.

The initial studies found evidence of early reliance on response learning. When infants were in an unmarked room (in which correct responding could only be based on dead reckoning), Acredolo (1978) reported that infants of 16 months were successful in turning to the correct side to see the event, but 6-month-olds and 11-month-olds continued the same movement they had made before (i.e., relied on response learning/sensorimotor coding). When a landmark (a yellow star) surrounded the correct window, and correct responding could be based on cue learning, 6-month-olds continued to make errors evidencing sensorimotor responding. However, about half the 11-month-olds showed correct responses. Some success at 11 months when a coincident landmark is available, but not otherwise, suggests an earlier advent of cue learning than of dead reckoning (at least, dead reckoning across the rotation-and-translation movement used in the study). Bremner and Bryant's (1977) data supported the hint in Acredolo's work that cue learning does not begin until toward the end of the first year. They found that at 9 months, infants continued to search as defined by response learning after having been moved to the opposite side of the table, even when cues marking the correct choice were provided by painting one side of the table black and the other side white.[1]

The findings reported by Acredolo and by Bremner and Bryant seem to support the existence of an egocentric to allocentric shift (although it might be better to use the term nonegocentric than allocentric, given that correct responding can be based on more than one kind of spatial coding). Specifically, infants in these studies seemed to search based on response learning, with some infants shifting to cue learning about 11 months, and with a shift to ability to use dead reckoning sometime before 16 months. However, there is another way to characterize development that does not involve wholesale shifts from one mode of coding to another. Various lines of evidence support this view, to which we now turn.

Questions about a Qualitative Shift

When organisms are stationary (and, of course, early in life, infant motor abilities are very limited), both response learning and cue learning lead to the same (correct) response, and dead reckoning is irrele-

1. The landmarks in these experiments surround the target: the yellow star surrounds the window where the event will occur, and the painted table surrounds the cloths hiding the toy. Are these landmarks coincident or distal? Their closeness to the target and the fact that they can be considered to touch it argue for considering them coincident, but clearly, the psychophysics (or Gestalt principles) of exactly what makes something coincident versus not remain to be worked out.

vant. But when organisms move, response learning is unreliable. Externally referenced systems and dead reckoning generally correspond after movement and either can be relied on when they do. In mature organisms any conflicts which do arise between external referencing and dead reckoning are usually resolved in favor of the externally referenced systems because, from a functional point of view, dead reckoning is more subject to drift and needs to be calibrated with respect to fixed external landmarks (Gallistel 1990).

Consider the possibility that infants must learn which spatial coding system to use when response learning conflicts with other systems, or when dead reckoning conflicts with externally referenced coding. Perhaps infants, with limited experience with the spatial world, have access to some or all of the possibilities for spatial coding but have an initial setting in which response learning predominates. Such a choice would not be maladaptive, since infants are initially fairly stationary. From an evolutionary point of view, after all, an organism whose initial reliance on response learning could be expected to be correct in any expectable physical environment would not have any problem surviving. All that is required for adaptation is that when response learning fails to locate an object, after active motoric abilities emerge, trial and error experiences in the world reveal that other coding systems would have led to the correct choice.

We will review several findings in the infant-to-the-opposite-side paradigm that support such a "reweighting" view of infant development. Infants show evidence of cue learning rather than response learning when the cues are fairly salient, as one would predict from the idea that their problem is deciding which of two coding systems to use when the systems conflict. Infants also show evidence of dead reckoning rather than response learning when the motions involved are actions that they have experience actively producing. This finding is again congruent with the idea of systems in conflict, since experience with certain kinds of motion is presumably crucial to determining what coding systems to rely on when. In addition there are other factors that apparently decrease the likelihood of relying on response learning, notably lack of recent reinforcement of such responding and/or lack of emotional stress. The existence of such factors undermines the idea of a qualitative shift, and supports the idea of coexisting systems whose relative importance is in flux.

Early Cue Learning

Recall that in Acredolo's initial work, 6-month-olds did not seem to use a coincident landmark (a yellow star around the correct window) to locate the interesting event, seemingly indicating an absence of cue learning; only some 11-month-olds used the yellow star to find the event. Similarly Bremner and Bryant (1977) found that painting half their table white and half black did not improve 9-month-olds' search for an object hidden under one of two cloths, one on the black-painted side and one on the white-painted side. However, the yellow stars and the white- and black-painted table tops turn out to be insufficiently salient cues to support cue learning. Bremner (1978b) showed that when the cloths hiding the objects were white or black, rather than just the table tops, 9-month-olds were able to choose correctly following movement to the opposite side of the table. Similarly Acredolo and Evans (1980) found that both 9- and 11-month-olds could make correct choices of where to look for an event when very salient landmarks (stripes and lights) surrounded the correct window. In fact, in the stripes-and-lights condition, infants as young as 6 months did not show definite signs of sensorimotor choices. Instead, they systematically checked both windows as if unsure about the location of the experimenter. This observation suggests that, even at 6 months, salient coincident landmarks are considered at least potentially relevant to spatial coding. The infants' vacillation also supports the idea that infants initially use both response learning and cue learning, but they have not yet worked out which system to rely on when they are in conflict. In sum, cue learning and response learning are both available at 6 months, and cue learning predominates over response learning by 9 months when the cues are more salient than those used in the initial studies.

Rieser (1979) reported further evidence of the ability of 6- month-olds to use cue learning under particular conditions, and of their lack of knowledge as to how to handle conflicts between the response and the cue systems. Infants in his study lay on their backs, looking up at four covers arranged in a diamond pattern, behind one of which an object was hidden. They were able to use a distinctive cover pattern to find a hidden object after they were moved in a 90-degree rotation on their axis, as long as the cover on the choice dictated by sensorimotor coding was unpatterned. When the covers of both choices were patterned, infants did not consistently use the contrasting pattern information to make the correct choice. However, in this condition, they did not show systematic sensorimotor choices either. Instead, they showed the same mixture of the two choices seen in the 6-month-olds studied by Acredolo and Evans (1980).

The conclusion that infants can use cue learning, that is, associate the location of an object with a coincident landmark, is reinforced by two studies in which coding systems were not placed in conflict. In a well-known study primarily directed at demonstrating early object permanence, Baillargeon (1986) habituated 6- and 8-month-olds to an

Habituation Event

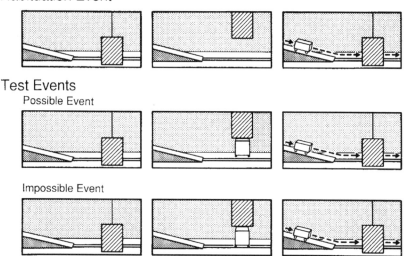

Test Events
Possible Event

Impossible Event

Figure 3.2
Infants are habituated to car rolling down track and behind a screen. They then see a block placed either behind the track (middle panel) or on the track (bottom panel). The question is how long will they look when the car rolls down the incline and emerges from behind the screen as usual. From Baillargeon (1993).

event in which a toy car went on a track down a ramp and then along a level plane in front of the infants (see figure 3.2). Following habituation, infants were shown a box placed either to one side of the tracks or on the tracks, and a screen was raised so that the infants could not see that part of the track. Infants who saw the box placed on the tracks looked longer when the toy car emerged from behind the screen (impossible event) than did infants who saw the box placed to one side of the track (either in front or behind). This study shows memory for the location of the box as on or off the track (as well as of the continued existence of the box, an issue in the object permanence debate). Thus the study strengthens the case for cue learning by 6 months, by showing that it is clearly seen in a situation in which there is no need to choose between systems in conflict.

More recently McDonough (1999) has shown that 7-month-olds can remember which of two distinctive containers contains an object, even after a minute's delay filled with distraction. Infants in this study were allowed no experience in reaching to containers, thus eliminating response learning in the motor system, and had their visual attention specifically attracted to both locations, thus minimizing response learning in the looking system. This study therefore shows clear evidence of cue learning at 7 months, at least in a situation where no choice among competing systems is required.

In sum, the data in these studies suggest that after 6 months infants code spatial location using coincident cues, as well as in sensorimotor terms. Which system they rely on when the systems are in conflict may be affected by stimulus salience, and by trial and error in finding which system works better.

Early Dead Reckoning

The movement infants underwent in the original work by Acredolo and by Bremner and Bryant was quite complex: a combination of rotation (spinning on one's axis) and translation (movement through space, here, on a semicircular path). Motion of this kind is unlikely to be experienced by nonlocomoting infants, except when being passively carried, a condition under which they generally pay little attention to their location (Acredolo, Adams, and Goodwyn 1984; Benson and Uzgiris 1985). Animal work in fact shows that spatially related hippocampal activity ceases almost entirely when active motion is restrained (Foster, Castro, and McNaughton 1989). However, fairly early in life, infants can engage in motion around their axis (i.e., in turning their heads and trunks from side to side, rolling over) and can produce simple lateral motion (i.e., by bodily swaying); they can also perceive their position relative to gravity (i.e., while lying down). Thus a natural prediction is that infants can take account of simple kinds of movement which they themselves produce, before they can deal with more complex kinds of movement which they cannot produce. That is, the initial studies may have underestimated infants' early abilities to engage in dead reckoning.

Several studies support this idea. At 5 months, infants distinguish their own motion along an arc around a stationary object from object motion around themselves while stationary, as well as from conjoint self and object motion (Kellman, Gleitman, and Spelke 1987). By 6 months, infants do not make egocentric choices when tilted with respect to gravity at training but when tested in an upright position (Rieser 1979), and infants can compensate for rotations to their left and right (Landau and Spelke 1988; Lepecq and Lafaite 1989; McKenzie, Day, and Ihsen 1984). By 9 months, infants can adjust for lateral movement along a line (Landau and Spelke 1988).

In sum, by 6 months, infants code spatial location using simple dead reckoning as long as they already have actively experienced producing the kinds of motion they need to take into account. At first, the motions they produce are limited, but being tilted with respect to gravity and head and trunk rotation do result in changes in apparent spatial location which babies can apparently take into account. Somewhat later, lateral translation can be added to this list, and sometime after that, more complicated paths involving both rotation and translation.

Initial Reinforcement of Response Learning
In Acredolo's and Bremner and Bryant's studies, the experimenters gave infants several trials of reinforced motoric experience before moving the infant. That is, the baby looked repeatedly, say to the left, and saw an adult making funny faces, or reached repeatedly, say to the right, and obtained an interesting toy. These trials seem likely to strengthen an infant's tendency to use response learning. While repetitions of certain motor sequences after infants are moved may reflect a conception of objects as at the disposal of action, as Piaget thought, it is also possible that more adaptive codings might be evident in situations in which they were not outweighed by the strength of the recently reinforced motor sequences.

One way to distinguish these alternatives is to examine performance following experiences in which infants see objects hidden but do not search for them. When prior motor movements have not been reinforced by obtaining an object or viewing a funny face, infants of 9 months are more likely to search correctly than when they have prior reinforced search experiences, although they do not entirely give up on egocentric responses (Acredolo 1979, 1982; Bremner 1978a). This improvement is evident in environments both with and without landmarks (Acredolo 1982).

Thus, at 9 months, infants may have some awareness that movement leads to a need to adjust location coding, and some ability to perform such adjustment, even after rotation and translation. Their ability to look appropriately—to choose the correct coding system from the conflicting possibilities—is reduced when one of the systems, namely response learning, has been recently reinforced.

Emotional Stress
There is another reason why the Acredolo (1979, 1982) and Bremner (1978a) studies just reviewed may have found better responding than had the initial studies using the paradigms. As well as not experiencing initial reinforcement of motor responses in these later studies, infants were tested at home, that is, in situations in which they likely felt less emotional stress than they may feel in typical laboratory situations. Emotional stress is known to affect spatial learning in rats (Diamond et al. 1996).

Curious whether the contrast between findings in laboratories and at home was due to emotional factors or to the fact that homes typically contain many more objects (and hence potential spatial cues) than do laboratories, Acredolo tested babies in an unfamiliar office environment—at least as cluttered as a home but still emotionally stressful, like a laboratory. She found babies making predominantly egocentric choices when tested in the office environment, strange like a lab but full of landmarks like a home. This finding suggests that negative emotion is more important in leading to egocentric choice in the lab than the absence of potential landmarks. Acredolo (1982) provided further support for the role of emotion, showing that even in a lab environment, 9-month-olds chose correctly rather than egocentrically, following movement to the opposite side of a room, if they were first given an opportunity to play in the environment and, presumably, to feel more comfortable in it. (At least they chose nonegocentrically when not given prior motor learning—no babies were tested in this study following initial reinforcement trials.)

Conclusion

The predominance of sensorimotor responding in Acredolo's and Bremner and Bryant's initial studies seems to depend on factors which affect how infants weight information when coding systems are placed in conflict. That is, infants' choices among conflicting spatial systems depend on some combination of cue salience, complexity of movement (and the associated question of whether or not the movement is already in the infant's repertoire), whether or not response learning has been recently reinforced, and whether the infant is emotionally secure or under stress. While not every combination of circumstances has been tested, it seems likely that at least by 6 months, infants can use both cue learning and dead reckoning with respect to self-produced actions, in addition to response learning.

It is interesting to realize that the early existence of cue learning was mentioned in Piaget's observations. Piaget's emphasis on action as defining location was combined with the suggestion that infants sometimes define location by association with landmarks. For instance, he wrote:

> Hence there would not be one chain, one doll, one watch, one ball, etc. . . . independent of the child's activity, that is of the special positions in which that activity takes place . . . but there would exist only images such as "ball-under-the-armchair," "doll-attached-to-the-hammock," "watch-under-a-cushion," "papa-at-his-window," etc. (Piaget 1954, p. 63).

Thus Piaget clearly acknowledged that infants might use both allocentric and egocentric learning (to use one set of terms) or both cue learning and response learning (to use another set of terms). However, he did not realize that infants might, early on, be able to correct self-referenced spatial coding using feedback regarding their own movement, when this movement is simple and self-produced.

Why Does Reweighting Occur?

The existence of infant capabilities in cue learning and simple dead reckoning, as well as in response learning, does not imply that there is no developmental change in spatial coding. On the contrary, there seems to be substantial change in the circumstances under which infants of different ages will and will not rely on response learning when such coding conflicts with cue learning or with dead reckoning. (There are also developmental changes in the accuracy and scope of dead reckoning, and appearance in the second year of place learning, matters we will discuss in chapter 4). According to the view we are advancing, development consists in changes in the importance attached to different types of spatial information as they come into conflict. Such changes would be represented as "reweighting" in a computational model. The changes may also turn out to have specifiable neural correlates.[2]

Changes in the cue validity of different types of information seem likely to be driven by success or failure in locating an object when relying on the various possible information sources. In this section we examine what changes in the capacities of infants and in the feedback they consequently obtain from the environment may drive a reweighting process.

Crawling

There are many dramatic moments in the infant's motor development in the first 18 months. The abilities to roll over, to sit up, to grasp, to pull oneself erect, and to walk transform the infant's experience. All might be expected to have some impact on the nature of the child's interaction with the spatial world. However, the motor milestone that has attracted the greatest attention from researchers interested in spatial location is crawling. Crawling represents the infant's first acquisition of independent mobility. Thus it has been suggested that it has

2 . Recent work with barn owls (Stryker 1999; Zheng and Knudsen 1999) is intriguing, although not directly relevant to spatial coding in humans. Young barn owls acquire auditory maps of space, stored in the inferior colliculus. These maps are calibrated with visual input. If visual input is changed by outfitting the birds with prism spectacles, a new auditory map is obtained. The old one is not, however, discarded, but is instead actively inhibited. It can be reactivated when the spectacles are removed though, even after the age at which plasticity is usually lost.

profound effects on the extent to which infants can keep track of their position in the world and of the location of objects following movement.

Several findings have provided support for this hypothesis. At 8 months, infants who can crawl show less sensorimotor responding in the Acredolo (1978) paradigm than noncrawlers (Bai and Bertenthal 1992; Bertenthal, Campos, and Barrett 1984). This difference is apparently not attributable to some maturational difference between the groups because noncrawling 8-month-olds with experience in walkers show less sensorimotor responding than did noncrawlers without walker experience (Bertenthal et al. 1984). In addition a case study of a child confined to a body cast for the first 8.5 months of life has shown that after release from the cast and the beginning of independent locomotion, the child showed an abrupt drop in sensorimotor choices (from 60 to 20 percent) in the Acredolo paradigm.

Some accounts of these findings stress the qualitative shift in functioning observed at this time (e.g., Acredolo 1988). However, if one thinks of infant spatial development as constituting a shift in how to weight coding systems when they are in conflict, the data on locomotory experience suggest, instead, that experience with particular kinds of movement helps to change the importance assigned to sensorimotor versus other coding systems, through positive and negative feedback. After all, it is only when infants move that response learning fails, providing information questioning its usefulness. Thus experience with gravitation (some of which is available even to the very young infant) is necessary for success in a task such as that of Rieser (1979), experience with rotation around one's trunk is crucial for success in rotational situations of the kind studied by McKenzie and Day or by Landau and Spelke, and experience with taking paths which summate rotation and translation, as occurs in crawling, is crucial for success in the Acredolo situation.

This view deemphasizes the idea of qualitative change from mode to mode of spatial coding, substituting a view in which particular kinds of motor experience affect the weighting of particular possibilities. From this point of view, it is quite interesting that development is piecemeal—that is, that experience with the failure of response learning following a certain kind of movement does not generalize to a deemphasis on response learning in all situations. Rather, infants appear to learn, situation by situation, what weights to attach to what kinds of information. Ultimately, of course, there may be a general reweighting (i.e., a developmental shift of default propensities) rather than more narrow, contextually based changes in how the system behaves in particular assemblies of circumstances. Distinguishing general reweighting from reweighting in every context likely to occur in the real world is a difficult matter, since clearly the two possibilities will usually produce the same adaptive behaviors in a wide range of circumstances. However, investigation of responses in novel circumstances might shed some light on the issue.

Visual Experience

Motor experience may not be the only experiential factor leading to a reweighting of possibilities. In particular, visual experience may also be relevant. The child confined to a body cast, studied by Bertenthal et al., showed an initial drop in sensorimotor responding (from 100 to 60 percent) while still confined to the cast, suggesting that it is not only locomotion that accounts for a diminution in sensorimotor coding. In addition, as already discussed in chapter 1, children who are blind in infancy are known to experience spatial difficulties, and children who have partial vision, or who had vision early and then lost it, have fewer spatial difficulties than infants profoundly blind since birth (Warren 1994), all of which suggests that visual experience contributes to spatial development, independent of locomotory experience. Blind infants often show delays in reaching motor milestones, but they do eventually crawl and walk. Yet, even after they acquire these motor abilities, their performance on many simple spatial tasks can still end up reduced.

Cortical Maturation

Another relevant factor in spatial development may be cortical maturation, although the nature of its role is not yet clear. Bell and Fox (1996) found that infants with more crawling experience had greater degrees of EEG coherence at frontal and occipital sites than infants with less crawling experience, and argued that crawling may reflect cortical maturation rather than exert an independent influence on cognition. A weakness of such an argument is that it would not explain the effect of walker experience found by Bertenthal et al. Another possibility is that the experiences occurring to the mobile child drive a process of brain maturation. Particular real-world experiences may lead to the adoption of certain task strategies that engage particular brain areas. Experience may even lead to structural change in the nervous system (as argued by, among others, Thelen and Smith 1994).

"Egocentric" Choices in Older Children

If one accepts the idea of a qualitative egocentric-to-allocentric shift, there is an apparently strange anomaly in development: egocentric choice among preschoolers in situations very similar to those in which infants succeed by 16 or 18 months. For instance, when 3-, 4-, and 5-year-olds are taught to play a game of searching for a prize, located under one of two buckets, either to the left or the right side of a small room (with the opposite bucket being empty), and then move to the opposite side of the room, only 5-year-olds consistently look for a toy in a way that allows for their changed position (Acredolo 1977). When reminded they had moved, 4-year-olds searched correctly. But 3-year-olds showed egocentric searches (i.e., response learning) even after being reminded of movement.

One might conclude that 3-year-olds are spatially egocentric. However, this seems an odd argument, considering that results in a very similar paradigm show egocentrism ending by 18 months (Acredolo, Adams, and Goodwyn 1984). However, on the view that spatial coding systems coexist throughout development, the data make more sense. Any spatial problem can be seen as requiring individuals to choose among representations that sometimes conflict. In the task given to preschoolers by Acredolo (1977), children are confronted with a socially defined game situation. Games, of course, can have arbitrary rules. The younger children are prone to favoring the quite plausible interpretation of the situation as a game in which you get a prize if you go to the left. (Experiments by Levinson 1996 show that this interpretation is made even by adults in certain situations.) Five-year-olds have what turns out to be a better first guess. This view of what is going on in that study emphasizes the fact that preschoolers differ from infants in their understanding of the socially situated nature of many cognitive tasks. Three-year-olds know this but infants don't. What 3-year-olds lack, and 5-year-olds have, is a good idea about the kind of game rule likely to be considered elegant by an adult.

An Earlier Transition: Retinocentric to Egocentric

Building spatial frames of reference is a protracted process. As we have just seen, older children sometimes show confusion about frames of reference. We should not neglect a brief look at children younger than 6 months as well. Infants of 6 months actually show a maturity, relative to younger infants, in their adoption of a body-centered frame of reference. There is evidence that at 4 months, visual orientation to the world is guided by retinocentric coordinates, in which visual position is coded in terms of where it appears on the retina (Gilmore and Johnson 1997). The shift from retinocentric to egocentric coordinates may reflect a transition from subcortical to cortical control, and maturation of areas of parietal cortex known to be responsible for combining information about retinal, head, and eye position to yield egocentric representations. In addition it may be that success and failure in fixating objects drives choice among these multiple competing systems and, in this case, preference for an integration of information from the systems. Thus in the first 6 months as in the second 6 months, development may consist in selection among multiple systems as motor control increases and as infants consequently have access to environmental information specifying adaptive functioning.

Section Summary

The evidence just reviewed suggests that, at least by 6 months, infants are able to code location using response learning (e.g., of head turns, eye movements, or arm movements), cue learning, and dead reckoning with respect to self-produced actions. However, in situations in which coding systems conflict, they rely on response learning more than do slightly older children, likely because they have insufficient experience to drive a determination of which system is more valid. Both motoric and visual experience appear necessary to "reweight" the system. Behavioral development is accompanied by changes in brain activity, but whether these are maturational, or the results of experiential input, remains to be decisively determined.

Overall, this way of thinking about early spatial development is quite similar to the model advocated by Siegler (1996) for analysis of development in general, and to the thinking of Hirsh-Pasek and Golinkoff (1996) about language development. The idea of variation-and-selection is clearly a powerful one in thinking about developmental change in more than one domain. In Chapter 8, we will discuss this theme more extensively.

What Does the A-not-B Error Tell Us about Coding of Spatial Locatation?

One of the central questions of this book is how we know where things are. Piaget asked a somewhat different question, namely how do infants know that there are things in the world at all. He studied infants' apparent knowledge of the location of things in an effort to answer his primary question regarding the existence of objects. The interleaving of efforts to address these two very different questions has led to a fair amount of confusion and controversy. In this section we review the controversy and the evidence, with a view to seeing what has been learned about spatial coding. We argue that analysis of spatial coding has been lacking in the A-not-B literature, which has tended to treat "memory for location" as if it were one self-evident thing, rather than a construct in need of analysis and likely standing for several different possible kinds of coding. If we think of "memory for location" in terms of possible ways of coding location, we argue that the findings on A-not-B errors can be interpreted as reflecting developmental changes in cue validities of various kinds of spatial coding in the first year, as already discussed with respect to the egocentric-to-allocentric literature.

Let us begin by reviewing Piaget's thinking on the matter. Piaget described infants as moving through six stages in their construction of the notion of what an object is, and where particular objects are, which for Piaget were completely intertwined notions. Initially infants will look for objects only when they are completely visible (Piaget's stages I and II in the development of the object concept); later a part of an object will trigger search (stage III). At stage IV infants exhibit the ability to search for objects that have disappeared completely. But they also show a curious error pattern. After having successfully searched for an object at one location (location A), they continue to search for it at that location even after observing the object being hidden at a new location (location B). Such errors are classically observed in infants of about 9 months, the age at which most research has concentrated (Wellman, Cross, and Bartsch 1987). At stage V infants have overcome the propensity for A-not-B errors but still cannot track objects if they are moved under a cover or in an experimenter's hand. They reach stage VI when they can follow a sequence of invisible displacements.

To Piaget and to many other readers, the A-not-B error of stage IV has seemed convincing evidence that infants initially consider objects as being at the disposal of particular motor actions. A second view, which has recently achieved fairly wide acceptance, is that infants have excellent spatial memory from quite early on, and that the A-not-B error therefore depends on additional factors, unrelated to spatial memory. Proposed factors include the ability to understand means-ends relations (Baillargeon 1993), to engage in problem solving (Aguiar, Kolstad, and Baillargeon, in press), or to inhibit a prepotent response tendency (Diamond 1990). Yet a third approach has emerged recently, critical of the competence-performance distinction which is central to viewpoints of the second kind (i.e., the argument that spatial memory exists but is masked by some additional factor or factors). One such critique comes from the perspective of dynamic systems theory (Thelen and Smith 1994), while the other arises from the tradition of connectionist modeling (Munakata et al. 1997; Munakata 1998a).

We examine current knowledge about the A-not-B error in four major steps. First, we summarize what is known about the A-not-B phenomenon by presenting a meta-analysis by Wellman et al. (1987), which reviewed what had been learned about the error up until ten years or so ago. We then discuss the subsequent investigations which modified two of Wellman et al.'s key conclusions. Second, we consider the evidence taken to support the view that young infants have good spatial memory for objects in the A-not-B situation, if memory is indexed by looking behaviors rather than by reaching. Third, many explanations of the A-not-B error refer to "memory for the correct location," but typically what kind of memory this might be has not been discussed in detail and has not made contact with the spatial coding literature. We attempt to rectify this lack. Finally we are in a position to discuss both the views of the error which rest on the competence-performance distinction and the views that challenge such an assumption.

Evidence on A-not-B Errors

Wellman, Cross, and Bartsch (1987) conducted a meta-analysis of the studies of the error available through that time. Their analysis concluded that A-not-B errors decrease with age, decrease when hiding locations are distinctive, increase with delay between hiding and infant search and, oddly, decrease with more hiding locations than two. Provocatively, they reported that the number of hidings at the A location did not seem to increase the proportion of A-not-B errors, a finding that seemed to pose problems for many explanations of the phenomenon.

The Wellman et al. analysis also showed that at each age studied, there were situations in which infants searched correctly, situations in which they showed the A-not-B error and situations in which they searched randomly. Just a few years before the meta-analysis, Diamond (1985) had reported a systematic relation between age, delay and likelihood of showing the A-not-B error. Infants as young as 6 months exhibit the error when delays between hiding and search are kept very brief, and infants as old as 12 months exhibit the error when delays are a bit longer than those traditionally used (Diamond 1985). Taken together, these two papers led to discomfort with Piaget's hypothesis that making the A-not-B error represented a distinctive stage in infant development—the phenomenon had proved discernible at a wider range of ages than one might have thought. The findings also led to a sense that there must be at least two forces at work to create the pattern of findings. One force would support correct performance but wane over the course of a few seconds, leaving a second tendency to produce A-not-B errors. Development could then consist in the appearance of longer-lasting or more robust bases for correct performance (e.g., better memory) and/or diminution in the force producing errors (e.g., better inhibition of prepotent responses).

The Wellman et al. work allowed investigators to focus their thinking about the phenomenon. Subsequent studies directed at the two most puzzling and troubling of the five conclusions have revealed that methodological problems accounted for both of them. The first puzzle was the apparent advantage of multiple-location over two-location situations, which Wellman et al. had recognized as having no obvious explanation. Diamond, Cruttenden, and Neiderman (1994) showed that the crucial issue was a seemingly trivial aspect of procedure. Experimenters using two wells could cover both simultaneously, using one hand to perform each action. But experimenters using multiple wells typically did not uncover and cover the nonhiding locations, but rather acted only on the hiding well. On the crucial trial at location B, they thus attracted visual attention to that area and enhanced its memorability, reducing reaches to location A. When a special apparatus was built that allowed for simultaneous covering of locations, even in multiple-well situations, performance was actually worse with multiple wells than with two wells—as had always seemed intuitively sensible.[3]

The second puzzle of the Wellman et al. monograph was the finding of no effect of number of hiding trials at location A. Many proposed explanations of the A-not-B error foundered on the fact that they predicted such an effect. However, Smith et al. (1999) have recently shown that, again, a seemingly trivial aspect of experimental procedure is crucial. Typical A-not-B paradigms include training trials designed to ensure that infants will search for hidden objects. These training trials are typically conducted at location A, and involve showing a desired object close to the box or well at A, then in the box or well, then partially covered. While intended only to induce babies to reach at all, such trials, Smith et al. argue, serve to build up response learning habits to asymptote, thus explaining the lack of effect of experimental trials at A. When they trained infants to reach using a different hiding location (location C) which was also a perceptually different box from the containers they used for their A and B locations, they indeed found a marked reduction in A-not-B errors compared to the standard proce-

3. Munakata (1998b) and Smith et al. (1999) do not agree that performance is worse with multiple locations. Smith et al. do not discuss the basis for their conclusion and their implicit dismissal of the Diamond et al. (1994) findings. Munakata argues, in response to Diamond et al., that Bjork and Cummings (1984) found an advantage for multiple wells even with simultaneous covering. But Bjork and Cummings do not state that they covered wells simultaneously in the Methods section of the experiment in which they compared a multiple-well and a two-well apparatus (Experiment 2). In the Methods section of their Experiment 1, using only multiple wells, Bjork and Cummings refer to covering the well in which the toy was placed but no other well, exactly the procedure that Diamond et al. argue leads to artifactually better performance with multiple wells.

dure. In addition Marcovitch and Zelazo (1998) have found, in a new meta-analysis of A-not-B studies, that there actually is evidence for an effect of number of A trials, contrary to the original Wellman et al. conclusion. Showing that prior reaching to location A does increase the likelihood of A-not-B errors removes a major roadblock from the task of understanding the error.

Location Knowledge as Indexed by Looking

Baillargeon (Baillargeon, DeVos, and Graber 1989; Baillargeon and Graber 1988; Baillargeon et al. 1990) studied memory for location in a situation in which objects were hidden behind two identical screens. An object was placed on one of two identical placemats (either to the left or the right of the infant), and the two screens were slid into position in front of the mats. A human hand appeared and, after a pause, reached behind either the left or the right screen and retrieved the object. When the object was found behind the wrong screen, 8-month-olds looked longer than when the object was retrieved from behind the correct screen, even when the delay between hiding and search was as long as 70 seconds. However, 7-month-olds did not show this effect, even with 15 second delays.

Baillargeon's findings on early location memory were immediately seen by many investigators as deepening the already difficult mystery of why 8-month-olds make the A-not-B error. It seemed paradoxical that infants should know the location of a hidden object perfectly well and yet reach for it in the wrong place. However, this reading of the Baillargeon studies was not strictly justified. The studies actually did not mimic the A-not-B situation: there were no initial trials in which the object was hidden behind one screen and appeared from behind that screen (location A), followed by a trial in which the object was hidden behind the other screen (location B). Thus the contrast between Baillargeon's findings and those in the traditional reaching paradigm could simply have been due to the fact that the traditional paradigm sets up conflict between response tendencies, whereas the Baillargeon experiment did not create such conflict.

Recently, though, there have been investigations of infant looking behavior which have included initial trials at one location, followed by a shift to hiding at a second location. These studies have provided evidence that looking times indeed show better memory for the location of a hidden object than does reaching behavior, even in an A-not-B analogue. Hofstadter and Reznick (1996) examined infants' gazes as well as their reaches in an A-not-B paradigm. Substantiating earlier anecdotal reports by Diamond (1985), Hofstadter and Reznick (1996) found that infants as young as 7 months were more likely to direct their gazes than their reaches toward the correct location (i.e., location B), and they did not show the A-not-B error in their direction of gaze. Interestingly, 5-month-olds (who were too young to reach) did show the A-not-B error in their gaze direction, showing that looking behavior is not immune from the error—it simply occurs in that system at an earlier age. Matthews, Ellis, and Nelson (1996) found in a longitudinal study, that the average delay needed to produce A-not-B errors at each age was about 2 seconds greater for looking behavior than for reaching behavior. Ahmed and Ruffman (1998) used a paradigm more analogous to that used by Baillargeon, in which looking times to possible and impossible events were monitored. They found that, even with initial A trials, 8- to 12-month-olds showed longer looking times to appearance at A than appearance at B following hiding at B (i.e., they seemed to expect the object at B, not at A, thus not committing the A-not-B error). These results appeared even following delays of 15 seconds, far longer than delays sufficient to produce A-not-B errors in grasping at these ages.[4] (Similar results have recently been found for the egocentric-to-allocentric shift paradigm reviewed in the first half of this chapter. Kaufman and Needham, in press, showed that 6.5-month-old infants show objective spatial coding as assessed by visual dishabituation, in the same paradigm in which Bremner and Bryant found egocentric reaching by infants as old as 9 months.)

The idea that motor behaviors such as reaching and visual attention are not always in congruence with each other has support in other lines of research than the A-not-B literature. Goodale and Milner (1992) have argued for this conclusion on the basis of work with brain-damaged patients. Support exists also in other lines of infancy research (Bertenthal 1996; Spelke, Vishton, and von Hoftsen 1995; Stulac and Vishton 1997). For instance, Stulac and Vishton found that 6-month-olds seemed to grasp for two objects when looking at visual displays in ways that

---

4. There is some controversy about the claim of better performance with looking than with reaching. Bell (1998) has reported that looking is not more advanced than reaching in an A-not-B paradigm. Matthews et al. themselves suggest that their 2 second difference may be simply due to the fact that it takes about 2 seconds for infants to initiate reaching. They argue that this delay means that the functional challenge to memory is longer in a reaching paradigm than a looking paradigm (i.e., the delay is longer). But this way of explaining the data seems a little odd. If an infant retrieves a memory when the display is revealed, that memory should inform both the oculomotor system and the grasp system. The delay to actually initiate a reach may be greater than the delay to move the eyes, but this seems irrelevant to memory. Once the intention is formed, time to initiate a reach should pose no additional burden to memory. Only delay during which nothing reminds the infant of the situation at all (e.g., when the infant is distracted from the display or when a screen hides the display) should affect memory.

suggested that they thought there was only one object. Specifically, they showed "separate object reaches" when looking at the objects with an occluded midsection used in a prior study by Kellman and Spelke (1983). Kellman and Spelke's investigation of looking times had suggested, in contrast to the results with reaching, that infants believed there was only a single object in these displays. While Smith et al. (1999) argue that looking and reaching are usually strongly coupled, this fact does not preclude the possibility that the two responses are of differential sensitivity in particular situations (Munakata 1998b).

In summary, although there are still puzzles to be worked out and some contradictory evidence, there is now some good reason to think that infants do seem to know more about spatial location than would be concluded from examining their reaching behavior alone. Exactly what they know is a matter of some controversy.

Memory for the Correct Location: What Kind?

Theorists attempting to understand the A-not-B error have often referred to infants' memory for the correct location (or lack thereof), without considering what kind of spatial coding might be involved in having such a memory. Consider what kind of spatial coding is required to remember which of two identical locations houses a hidden object. Cue learning is insufficient, because the screens are identical. (When they are not, or at least when they are different in a salient way, infants' ability to reach correctly is enhanced; e.g., Diedrich and Highlands 1998). Externally referenced coding using distal landmarks is unlikely because such landmarks do not seem to be used until late in the second year of life (DeLoache and Brown 1983; Mangan et al. 1994; Newcombe et al. 1998; see discussion in chapter 4). Dead reckoning is irrelevant because the infants do not move.

One likely kind of spatial coding is the kind we have already seen is favored by infants: response learning. Infants can remember their looking (e.g., in terms of head orientation or gaze direction); specifically, infants of 6 months have been shown to be able to retain information to guide looking behavior in spatial working memory for at least 5 seconds (Gilmore and Johnson 1995). In addition it seems likely that infants can retain a working memory of their reaching (e.g., in terms of shoulder orientation and arm extension).[5] Monkeys have

5. Response learning might be aided by having "place holders" available in the two identical containers or cloths; such markers allow for a categorical choice between two different head orientations or reaching directions, and eliminate any necessity for coding distance in a fine-grained way.

been shown to have neurons firing in ventral premotor cortex when they see stationary three-dimensional objects, which continue to fire even after the lights are turned off and the object is silently removed (Graziano, Hu, and Gross 1997). Rizzolatti et al. (1997) argue that this firing most likely represents the continued activation of motor schemata.

In terms of looking responses, the A-not-B situation requires infants to choose between the most frequently executed look (looking to A) and the most recently executed look (looking to B). In terms of reaching responses, the A-not-B situation requires infants to choose between the most frequent and the most recently executed reach (reaching to A) and a less frequent and less recent reach (reaching to B)—a contest obviously usually won by reaching to A. Thus reaching for the toy at its current location is inherently more complex than looking at the toy in its current location, in the sense that infants need to base their reach not on response learning in the reach system but on the most recent looking response (looking to B).

There is another kind of spatial coding possible in most A-not-B experiments, which might be used to guide either looks or reaches. Infants have recently been shown to have some capacity to code location in continuous space, with reference either to a surrounding geometric frame or to their own coordinates (Newcombe, Huttenlocher, and Learmonth, in press). In most experiments the A and B hiding places are presented with some kind of surround, such as the edges of a table, the frame of an apparatus that allows a screen to descend, and with hiding places in a constant position with respect to the infant's body. Thus it is likely that infants can code the location of the A and the B locations with respect to the surrounding frame and/or with respect to distance/angle from their own position. (Such coding is probably different from coding with respect to distal landmarks, which we have said is apparent only in the second year; see chapter 4 for discussion.) Using coding of spatial position in continuous space, in preference to an incorrect response memory and/or as a supplement to a response memory for the most recent look, could guide correct looks or correct reaches. Postulating such information explains the existence of spatial gradients in infant search, in which incorrect searches are biased toward the correct location (Bjork and Cummings 1984; Diamond et al. 1994).

In summary, "memory for the correct location," a phrase often used in A-not-B studies but rarely defined, is likely to be one of two kinds of memory, perhaps both: memory for the most recent looking response, or memory for the distance of an object from surrounding frames available in the environment (e.g., the edges of a table, or the infant's own position). Responses to the incorrect location may also be based on spatial memories. Memory for the incorrect location could consist of response memories for frequently executed responses in the visual system, and/or response memories for both most frequent and most recent responses in the reaching system. Viewed in this way, the conflict that results in the A-not-B error is quite similar to the conflicts between coding systems that we analyzed when examining the egocentric-to-allocentric shift. However, most of the conflicts occur within the response learning system.

Some evidence in support of this way of conceptualizing the spatial-memory demands of the A-not-B task comes from the fact that locomotor experience affects performance on A not B, just as it affects performance on paradigms in which the infant moves (Bell and Fox, 1997; Horobin and Acredolo 1986; Kermoian and Campos 1988). This effect seems initially a little puzzling, in that the infant does not move when performing object permanence tasks. However, at the most general level, the effect of movement through the world can be conceptualized as teaching the infant the lesson that some response memories are unreliable. In particular, they could be expected to learn: (1) that recency is more important than frequency, (2) that the most recent response of any kind, be it looking or reaching, which was associated with the object, is to be preferred to the most recent response judged only within a single system, and (3) that codings of spatial position not using response memories at all, but rather using external frameworks, are particularly reliable. Reduction in reliance on frequent responses or most recent reaching responses, and increases in reliance on most recent looking responses or external frameworks, would increase the likelihood of correct responding on the A-not-B task.

Accounts of A-not-B in Which Infants Are Assumed to Know Location
Recent findings on infants' abilities to locate objects by looking demand explanation in any approach to A-not-B errors. An obvious way to approach the problem is to suggest that infants' looking shows that they know the correct location and to explain incorrect reaches as the product of some second factor. Several such accounts have been proposed. These approaches are distinguishable ones, but from the point of view of understanding spatial development, they share an important element: they take as given that infants may remember the spatial location of the object even as they commit the A-not-B error.

Means-Ends Theories
One possible explanation for the contrast between findings on infants' knowledge of location as indexed by reaching in the A-not-B task and

infants' knowledge of location as indexed in looking time experiments has been that reaching requires the coordination of several steps to reach an end, whereas looking does not require such means-ends coordination (e.g., Baillargeon 1993). Means-ends reasoning develops over the infancy period and, as indexed by traditional Piagetian tasks, does not reach a mature form until the middle of the second year (Gopnik and Meltzoff 1987; Uzgiris and Hunt 1975).

However, three findings argue against a means-ends analysis. First, when Hofstadter and Reznick (1996) compared the correctness of direction of gaze with the correctness of direction of reach, they credited infants with correct reaches even when they only touched the correct location. That is, infants did not have to move the cloth and obtain the object in order to be counted as correct reachers, thus eliminating the means-ends component of reaching. Direction of gaze was more likely to be correct than reaching, even with this kind of scoring. Second, Munakata et al. (1997) examined 7-month-old infants' ability either to pull a towel or to push a button to retrieve an object. They found that the infants had no trouble with using such means to obtain a toy as long as the toy was visible; difficulties arose when the object was hidden. They argue that this observation rules out means-ends analysis as the source of difficulty in infant search; using a means to obtain an end can be accomplished as long as that end (i.e., the object) need not be mentally represented. Third, although Willats (1999) criticized the Munakata et al. study on the basis of his finding that 7-month-olds were only beginning to show intentional means-ends behavior in the towel-pull situation, at the distances used in the Munakata et al. study, another aspect of his study also weakens the means-end position. He found that 8-month-olds showed very clear intentional pulling, even at far distances. However, infants of this age would clearly be still committing the A-not-B error. Fourth, as might be predicted from point 3, a longitudinal study of term and preterm infants found no relation between performance on the A-not-B task and performance on a means-ends task involving pulling a cloth to retrieve an object (Matthews, Ellis, and Nelson 1996).

Problem-Solving Theories
Difficulties with means-end analysis do not preclude the possibility of accounting for A-not-B errors using a model in which infants know location but do not always display their knowledge. Aguiar, Kolstad, and Baillargeon (in press) propose a problem-solving approach. In this way of thinking, perseveration is the key issue. These investigators remind us that we can see perseveration in tasks in which no location memory is required. For instance, infants will pull on a cloth to obtain a toy, but they sometimes will pull even when the toy is not attached

to the cloth. In this case one would say they simply don't understand the problem. They haven't appreciated the importance of attachment, or haven't noticed that the object was unattached. When these matters are highlighted, and infants realize the importance of attachment and when the toy is attached and when it is not, perseveration in the not-attached situation was said to diminish.[6]

As applied to the A-not-B paradigm, a problem-solving analysis suggest that at a certain age infants do not appreciate that a change in the hiding place requires a change in their response. This model makes predictions about A-not-B errors that are interesting and need to be evaluated. In particular, a problem-solving analysis would predict that manipulations designed to highlight the crucial aspects of the A-not-B situation would result in within-age improvement in performance. That is, unlike a model in which response inhibition needs to mature or working memory needs to strengthen, the problem-solving model predicts trainability of the A-not-B task.

Memory/Response Inhibition Theories
Diamond (1985, 1990, 1991) considers two factors as determining infant performance in the A-not-B situation: memory for the most recent hiding (at location B), and habit of reaching to location A. Memory for hiding at location B is considered to be strong initially but to wane over time. At the same time the infant has a tendency to reach to location A, based on prior experience with the object at A. Thus this model predicts that at short delays infants reach correctly, but as delays lengthen, they show the A-not-B error. With development, memory becomes stronger and more persistent over time, and habit is weakened by the onset of inhibition, made possible by maturation of prefrontal cortex. Thus the delay at which the A-not-B error is found becomes longer. Eventually it is not seen at all.

The evidence for this model is multifaceted. One part of the proposal is that infants have a memory for the current spatial location of the object, although a memory that, like all memories, weakens over time. This postulate is supported by the fact that the length of delay that an infant can tolerate before making the error increases in a highly regular manner (Diamond 1985). Other evidence for a memory component involves the fact that errors in multiple-well tasks are not random but cluster around the correct well (Bjork and Cummings 1984).

Another part of Diamond's proposal is that inhibition of the response of reaching to the habitual location is required to prevent the A-not-B error. This idea is supported by the fact that errors in a multiple-well

6. However, this matter will need further examination, because Willats (1999) found no difference between attached and nonattached conditions.

experiment are not randomly distributed around the correct well, as a pure memory explanation would predict, but instead are more common on the side closest to the previously correct location, location A, as if there were a "pull" to that location (Diamond et al. 1994).[7] Additional support for the idea that inhibition develops over the 6- to 12-month age range comes from situations in which infants are observed attempting to retrieve objects from a Plexiglas box open on only one side. If the open side faces infants, they can retrieve the object, but when the open side is not directly in front of infants, they initially have trouble getting an object. They reach for it directly, hitting the covered side. Arguably, this action reflects a difficulty to inhibit the prepotent response of reaching in a fashion determined by line of sight.

The third distinguishable aspect of Diamond's approach is the suggestion that inhibition of a prepotent motor response depends on maturation of prefrontal cortex. Animal studies have shown that lesions to dorsolateral prefrontal cortex, but not to parietal or hippocampal cortex, produce A-not-B errors and also lead to failure on the object retrieval task (Diamond 1991). In addition electrophysiological recordings with human infants link development of A-not-B skills to maturation of dorsolateral prefrontal cortex (Bell and Fox 1992, 1997). Maturation of dorsolateral prefrontal cortex continues over the first five to six years, as indexed by behavioral measures (Gerstadt, Hong, and Diamond 1994) and by neuroanatomical studies (P. Huttenlocher 1979). Thus it is also interesting to note that perseverative errors in spatial search tasks are not restricted to infants but extend across the first years of life. For instance, Perlmutter et al. (1981) found perseveration in nine-choice search tasks in children as old as three years, and Spencer, Smith and Thelen (1997) found that 2- and 3-year-old children perseverated in searching in a sandbox task in which, they argued, the absence of visual cues makes the bias induced by prior reaches a strong force on behavior.

In summary, Diamond's account of the A-not-B error suggests that infants' ability to succeed in the task depends largely on maturation of prefrontal cortex sufficient to allow for the inhibition of a prepotent response (i.e., reaching in a certain direction as determined by habit) and reliance instead on "memory" for the correct location. Diamond does not analyze what the nature of this memory might be, but we

7. However, an alternative explanation of the Diamond et al. data is that they simply show expected effects in spatial memory. Apparently, a "pull" toward A in their experimental setup would also be a bias toward the center of the display. Such bias is predicted by the hierarchical coding theory discussed in chapter 2, and it has been shown to exist in children at least as young as 16 months (Huttenlocher, Newcombe, and Sandberg 1994; see also chapter 4).

have already argued that it could be either memory for the most recently executed response in the looking system (Gilmore and Johnson 1995), or memory for location in continuous space with respect to a surrounding reference frame such as the edges of a table (Newcombe et al., in press).

There are, however, problems with one aspect of Diamond's hypothesis, namely the causal role ascribed to frontal lobe maturation. As we argued with respect to the egocentric-to-allocentric hypothesis, various spatial coding rules appear to coexist in early development. Feedback from environmental success and failure is required to choose among them or to weight them appropriately. Prefrontal cortex might be the locus for such weighting, since instantiating weighting may entail inhibition as well as strengthening of particular coding rules. In this case one can imagine that physiological maturation might not be the *cause* of behavioral change but rather the locus of its realization.

One hint of the promise of such an interpretation comes from Hofstadter and Reznick (1996). Recall that they found A-not-B errors in gaze direction among 5-month-olds, errors that had disappeared by 7 months. Children have been looking since birth, and looking with some degree of head and neck control from the age of a month or two, but they have only been reaching for objects from 4 or 5 months on, and even then are still working on their reaches. Thus one natural way to explain the "decalage" in when the error appears is to suggest that a few months of feedback from the environment about how to weight amount versus recency of response-based codings is required to settle on a recency-based rule. Inhibition of amount-based responding would be executed in prefrontal cortex. This reinterpretation is similar to that advocated by dynamic systems theory, to which we now turn.

Accounts of A-not-B in Which Infants Do Not "Know" Locations

Not all investigators have taken the finding that infants look better than they reach to indicate that a competence to locate objects is present early, but simply masked by performance factors such as means-ends reasoning or inability to inhibit prepotent response tendencies. There are two recent accounts of the A-not-B error that question this application of the competence-performance distinction. In one account, infants do or do not make the error as a function of the current state of inertial response tendencies set up by recent histories of activity in the looking and the reaching systems (Smith et al. 1999; Thelen and Smith 1994). In a second account of the A-not-B error, infants' object concepts are said to be fragile early on, and to strengthen over time and experience. More consolidated concepts are needed to support the reaching behavior observed in the A-not-B paradigm than to support the visual responses taken to indicate early spatial memory (Munakata et al. 1997; Munakata 1998a, b).

Dynamic Systems Theory
Thelen and Smith (1994) suggest that attention in the visual and manual systems may conflict during the A-not-B sequence. The manual system is more likely to show A-not-B errors than the visual system because it is more affected than the visual system by inertial properties set up by the initial hidings at A. The group headed by Thelen and Smith have discovered several phenomena consistent with a dynamic-systems analysis. First, as we have seen, Smith et al. (1999) showed that, contrary to the findings of the Wellman et al. (1987) meta-analysis, likelihood of the A-not-B error does depend on the number of recent reaches to the A location, as it must according to their inertial properties argument. Second, the error does not depend on an object being hidden at all. Smith et al. (1999) reported that infants will make the A-not-B error even when there is no hidden object, reaching for the lid on an empty box which has been more frequently (but not most recently) waved by the experimenter. Third, attracting infant attention to a certain side increases the likelihood that the infant will reach to that side, showing that performance in this situation has as much to do with task dynamics as with knowledge of location (Smith et al. 1999). Fourth, the likelihood of making an A-not-B error is reduced when infants have weights attached to their arms (Diedrich 1997; Diedrich, Thelen, and Smith 1998). One way to think of why weights have this effect is to suggest that the weights disrupt the task dynamics of reaching. An alternate phrasing of this idea might be that the heaviness of the arm reduces the attractiveness of relying on response learning, and prompts a reconsideration of the (mental) weights to be attached to the various competing coding options. Fifth, when infants are moved from a sitting to a standing posture between A trials and B trials, A-not-B errors are virtually eliminated (Smith et al. 1999).

Another recent paper from the Thelen and Smith group analyzed the object retrieval task which Diamond claimed also showed the difficulty infants have inhibiting a prepotent response, in this case, reaching on a direct rather than circuitous path for a visible object. Titzer, Thelen, and Smith (1995) showed that infants' ability to retrieve objects from boxes with openings facing sideways is heavily influenced by experience. Most infants have never faced this situation before being seen in experiments, but practice with such boxes improves their performance considerably. Further, operating on transparent boxes is influenced by experience with transparent solids, which are not often encountered by young infants. In other words, experience with relevant situations affected the object retrieval task markedly; maturation did not seem to be influential.

There are three crucial differences between the proposals of Diamond and of Smith and Thelen and their co-authors however. One difference is that Thelen and Smith are skeptical about maturation. Thelen and Smith find the direction of causation advocated by Diamond simplistic. They believe that changes in the dynamic equilibrium of the systems in conflict in the A-not-B situation can account for change at the behavioral level. These changes in dynamic equilibrium appear to be related to locomotor development. Locomotor experience may drive changes in the neural substrate, in this case in the dorsolateral prefrontal cortex. Thus, behavioral changes are implemented at the neural level, but need not be caused by maturational changes at the neural level.

The second difference between the two approaches is their treatment of the competence-performance issue. In the explanation advanced by Diamond, infants within the first year of life are thought to code the spatial location of objects (i.e., to have the competence to remember location), while having difficulties of various kinds in acting on this knowledge in certain conflictual situations. Thelen and Smith strongly reject the competence-performance distinction, writing that "knowing is what infants do in both looking for and searching for hidden objects. Knowing is [a] process of dynamic assembly. . . . We do not need to invoke represented constructs. . . . Logical structures for these constructs do not exist outside the task that invokes action." (p. 310) We will say more about this issue shortly.

The third difference is in the postulation of inhibition. Inhibition is a key factor in Diamond's model. But the Thelen-Smith position is that nothing needs to be inhibited for more mature performance to emerge; rather, there is simply a shifting balance of forces in the dynamics-determining behavior.

Connectionist Models

Munakata et al. (1997), as already discussed, conducted several experiments arguing against a means-ends account of infant successes with looking and failures with reaching in object search tasks. These investigators follow up this empirical work by arguing against accounts of development in which early success implies knowledge of principles and any failures are attributed to performance factors (i.e., they argue against the competence-performance distinction also criticized by Thelen and Smith). Using connectionist modeling techniques, they offer instead an "adaptive processing" account in which knowledge is graded.

In the Munakata et al. simulation, the representational system learns gradually to represent occluded objects, through repeated exposure to situations of occlusion and reappearance. The differences between predictions made by the system about what should be in the input are compared to the actual situation to drive learning. Two output systems are also included in the model, both dependent on a single representational system. One system is the perceptual prediction system, while the other is meant to stand for manual responses. Munakata et al. observe that reaching requires greater effort and occurs less frequently than looking, and additionally that infants fixate interesting objects from birth but only reach for objects beginning at about 4 months. To simulate these observations, Munakata et al. set the manual output system to begin learning after the perceptual prediction system had partially learned about occluded objects, and by setting the manual system to learn at a tenth of the rate of the perceptual prediction system.

Munakata (1998a) has supplemented this model with a model specifically addressing the A-not-B error. In this model responses can be based either on active memory traces, which develop, or on latent memory traces, which tend to support perseveration. This model is shown to be capable of simulating several specific findings in the A-not-B literature. It also makes novel predictions, whose testing will provide critical feedback about the virtues of the approach.

The connectionist models advocated by Munakata et al. (1997) and Munakata (1998a, b) are interesting, and the work constitutes an existence proof that the infants' behavior can be produced by a being with this kind of cognitive architecture. However, there is no direct evidence, as Munakata et al. acknowledge, that the models capture the way infants actually learn (a point also made by Mandler, in press). For instance, a competing connectionist model of object permanence uses very different assumptions (Mareschal, Plunkett, and Harris 1995).

Synthesis

One way to think about the A-not-B situation is that it requires infants to choose between two locations, each with some evidence in its favor. Location A is the most frequently looked-at and most frequently reached-to site, as well as the one most recently reached to. Location B is the most recently looked-at site, as well as the one whose location may be known as that of the desired object, in terms of a frame of reference such as the edges of a table or the baby's own body. Choosing among these various possibilities for basing a response requires knowledge of the cue validity of these pieces of information, knowledge that may require experience with the consequences of each over the course of a few months. That is, appropriate responding depends on assessing the relevance and importance of conflicting pieces of available information, as already stressed in the discussion of the waning of egocentric responding. A particular problem for infants faced with this reweighting task may be that the tendency to use frequency is initially so strong that it must be actively inhibited. Executing such inhibition may engage areas of prefrontal cortex, a function that could be conceptualized as tuned by experiential input over the course of several months. An additional factor, which could account for these findings instead of or in addition to response inhibition, might be the strengthening of spatial working memory. Interestingly the locus of spatial working memory is also prefrontal cortex (e.g., Jonides et al. 1993). The studies of spatial working memory have involved the location of a symbol on a card, so the spatial coding assessed could either involve most recent looking direction or a coding of coordinates with respect to the frame of the card.

Looking and Reaching

Why is looking more advanced than reaching? A simple, but perhaps correct, possibility is that grasping simply starts later than looking. Saying this entails accepting the idea that development in the two systems may, at least initially, be rather separate (a theme Piaget would certainly have endorsed). Additionally there is more to inhibit in the case of reaching. The most recently executed looking response is usually to the correct location (and when it is not, as Smith et al. have shown, A-not-B errors go up). By contrast, the most recently executed reaching response is usually to the incorrect location.

Competence and Representation

When do infants "know" where the toy is? Some theorists have approached this issue by saying that the earliest appearance of success shows the advent of competence, while others have denied that the concept of competence is a meaningful one at all. Neither option is attractive. The first position involves undue emphasis on behaviors that are restricted in scope. When a developing organism can only succeed in doing something in a situation carefully tailored to be as simple as possible, that success may not predict any functionally useful ability and is unlikely to allow for adaptive action in the real world. The second position devolves quickly into an uncomfortable degree of specificity. The aim of psychological analysis is to make generalizations across as many task contexts as possible, consistent with observing distinctions compelled by real behavioral phenomena.

Fortunately there is another possible approach to the problem of defining competence. In this way of thinking, any observed ability acquires meaning only as set in a developmental context (cf. Fischer and Bidell 1991). Certain abilities develop in a certain sequence. Each ability in a sequence is generally more adaptive than its predecessor—more general, more powerful, more accurate—while at the same time having limitations that will be overcome at a later time in development. Mature competence is, then, whatever point human beings generally reach when a developmental sequence ceases to advance. Generally, this state will be at least tolerably adaptive and will include the coordination of different neural systems involved with the same domain. This mature competence may be variable among individuals, across situations, and through historical time, given variations in environment, although certain essential aspects are likely to be universally evident due to selection pressure.

In this view, defining competence is equivalent to setting a criterion. Theorists must decide where along a developmental progression something interesting has happened. Clearly, what is "interesting" is going to be relative. For certain analytic purposes, the criterion of competence could be the first tentative outcropping of an ability. For other purposes, it may be the robust demonstration of this ability in every instance, even in the most demanding situations. In other cases, it may be something in between.

The setting of a criterion implies the existence of a certain kind of "competence." Thelen and Smith object fiercely to this term, but it is compatible with dynamic-systems theory in the usage proposed here. The very young infant does not have walking competence, in this view, because the circumstances in which walking can be executed are so painfully limited. And adults do have walking competence because they can walk in all the circumstances in which they usually find themselves, even though most of them could not walk over tightropes and none of them could walk on a planet with triple the earth's gravitational pull. Similarly competence defined as criterion setting is not incompatible with the connectionist models advocated by Munakata et al. (1997) and Munakata (1998a, b).

Summary

In this chapter we have argued against the view that there are various spatial coding systems that succeed one another in the course of infant development, and, in particular, against the view that an egocentric sensorimotor creature is changed by the onset of locomotion into a creature engaging in allocentric spatial coding. There is evidence that at least by 6 months infants code location using cue learning as well as in sensorimotor terms, and that they take into account the results of simple kinds of movement with which they have experience. Developmental change consists in changes in the importance assigned to these systems when they are placed in conflict, a process that likely occurs both because of experience with varied kinds of movement and because of visual experience coordinated with movement. Such an analysis is not foreign to the spirit of much of Piaget's treatment of infancy, but it differs substantially in detail and emphasis.

The second topic of this chapter was the A-not-B error. We have argued that the error, like egocentric choices following movement, involves a conflict between various modes of location coding. Resolving the conflict requires knowledge of the cue validity of the various coding possibilities, knowledge that may require feedback from the environment over the course of a few months. That is, appropriate responding depends on changes in the importance given to various types of available information. These changes may, at least at certain developmental points, require active inhibition of a dominant response; they may also depend on strengthening of codings more likely to lead to successful task performance, specifically, memory for the most recent look and memory with respect to external coordinates. Both inhibition and memory strengthening may themselves depend on environmental feedback and be implemented in frontal cortex, although an alternative is that cortical maturation is required. Solving the A-not-B problem is then a matter of working out the adaptiveness of various locational coding systems and coordinating various response systems, notably vision and motor response. Mature competence consists in the state of optimal weighting of spatial coding systems, coordinated across response systems. Initial competence consists in the possession of the key coding options.

Chapter 4

Three Other Important Questions about the Development of Location Coding (and an Epilogue on Automaticity)

In chapter 3 we reviewed research examining the Piagetian hypothesis that infants begin the task of spatial development equipped with only one of the four forms of location coding described in chapter 2, namely response learning. We argued that the evidence supports a somewhat different conceptualization, one in which response learning coexists during infancy with two other forms of location coding, namely cue learning and simple forms of dead reckoning. A central aspect of development in infancy seems to be evolution in how these systems are weighted in situations of conflict. This reweighting occurs in response to experiences arising during various kinds of movement in the world, as well as during visual experience.

While this is a good beginning for understanding the early foundations of spatial coding, it fails to deal with many aspects of mature spatial coding. Place learning is ignored. This is a major omission since coding distance with respect to distal external referents is a very flexible and powerful system, one that is the final arbiter of location for mature organisms in cases where coding systems conflict (Mittelstaedt and Mittelstaedt 1980). Another vital aspect of mature spatial functioning, hierarchical coding, is also ignored, even though the adaptive combination of various kinds and grains of spatial information across the inevitably embedded nature of the spatial world is essential to spatial competence. Last, although chapter 3 dealt with some aspects of dead reckoning, powerful abilities to integrate distance and direction over extended paths of movement in the environment continue to develop beyond infancy.

In this chapter we attempt to remedy these omissions. In addition, in an epilogue, we address the claim that spatial coding is automatic rather than effortful and does not change developmentally (Hasher and Zacks 1979). This point of view seems to conflict with all the discussion of development of spatial coding in chapters 3 and 4 of this book; we try to reconcile the ideas that spatial coding is automatic and that spatial coding develops, by proposing an automatic core nonetheless supplemented in very important ways during development.

Development of Place Learning

As we argued in chapter 2, the surest knowledge of the spatial location of an object comes from coding its location with respect to available landmarks in the world, using sets of distance and angle information anchored on a common set of landmarks, sometimes called a *frame of reference*. If enough information is remembered to specify location uniquely, then as long as those landmarks are perceptually available (or their location can be inferred), then the location of the target object is known. Movement about the world, with its attendant danger of gradually increasing error, can be safely ignored or used as a complement to externally referenced coding.

There appear to be two separable aspects to coding location in terms of distal landmarks. One is the ability to code distance in continuous space at all. This ability is necessary but not sufficient for the use of distal landmarks as reference points for such distance coding, or place learning, which entails also having the ability to coordinate distances coded from several landmarks often far removed from the target location.

Coding Distance in Continuous Space

For several decades, following Piaget's characterization of children's spatial representation as topological, researchers assumed that infants and toddlers had very little, if any, ability to code spatial extent or distance. Belief in this assumption had a profound influence on interpretation of data. For instance, to look at one example in some detail, Allen (1981) concluded that children as old as second graders code space only in categorical terms, and that they do not use distance. He based this conclusion on a study of second graders', fifth graders', and adults' ability to select which of two locations was closer to a third (reference) location. His interpretation of the data ignored the fact that second graders showed substantial, if not quite significant, effects of distance in their abilities to solve these problems (so that insufficient power to detect an effect may account for the null finding). But the "topological space" conclusion was so attractive at that time that most readers were happy to accept it.

Recent investigations have, however, challenged the twin ideas that children cannot code distance and that their spatial representations are topological. In recent research we have found that even toddlers seem able to notice distance and use that information to find objects. Emboldened by the robust competence found as young as 16 months, we have worked backward in developmental time to examine the earlier origins of coding of extent in infancy.

Toddlers

There are two conditions for studying coding of extent in very small children. First, performance in the experimental paradigm must index extent. While this is not only obvious but tautological, much research on spatial coding using infants and toddlers has centrally concerned other questions than whether they code distance. As a consequence the research has involved situations where an object can be located in one of two possible places. These possible locations are often identically marked: the same container, handkerchief, or window frame define the areas for search. In such tasks, one is not asked to code distance in continuous space, only to distinguish two places, for instance, in terms of left versus right of the body midline. Even when the number of possible places is expanded beyond two, in order to test various hypotheses about infant search (e.g., Bjork and Cummings 1984; Sophian 1984), it is not clear that one is assessing children's ability to code distance in continuous space; the space presented to the child is still punctate rather than continuous. Thus, in studying the early coding of extent, it is essential to use a continuous space. A second necessary criterion for studying extent with very young children is that one wants to examine coding in a situation as simple as possible.

To accomplish these two goals, Huttenlocher, Newcombe, and Sandberg (1994) used a sandbox, five feet long and one foot wide, in which children observed objects hidden under the sand (see figure 4.1). Once the sand was smoothed over, children saw a continuous space. We then asked them to indicate where the object was located, within that continuous space, by attempting to find it. One object (a small attractive toy) was hidden at a time, at one of nine predetermined locations at six-inch intervals in the sandbox. Children were allowed to search for the toy after turning to give their parent a kiss (to break eye gaze or motor straining toward the correct location), an action that took only a few seconds. Thus we used a very simple situation in which extent needed to be coded along only one dimension, for one object, for a short delay period, in order to do something quite straightforward, namely find the object. (The requirement to code along only one dimension was met because, although the sandbox was one foot across, objects were always hidden in the middle of the width, that is, six inches from either side so that variation along this second dimension was irrelevant to finding the object.)

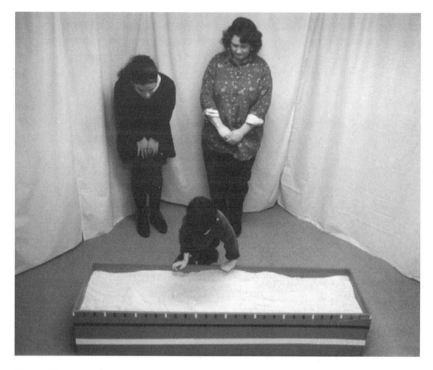

Figure 4.1
A toddler searches for an object hidden in a long rectangular sandbox. Procedure used in Huttenlocher, Newcombe, and Sandberg (1994).

The findings were quite clear. Children as young as 16 months were able to remember the location of objects hidden, one at a time, at one of nine locations along the five-foot length of a sandbox. Figure 4.2 shows their average search positions plotted against the correct search positions, and there is obviously a very close correspondence between the two. Statistically the children showed highly significant differentiation among the nine hiding locations, confirming the visual impression made by the graph. Furthermore this accuracy was as clear at 16 months as at 24 months, with no developmental change found across the 8-month time span, and accuracy remained high even when children moved laterally along the box before being allowed to search. Thus children are apparently capable of coding extent in a fairly precise way by 16 months, using as their referent the frame of the box, rather than their own position.

Even earlier coding of extent, and in a two-dimensional situation, was observed by Bushnell et al. (1995). These investigators observed 12-month-olds searching for objects hidden under one of 58 identical

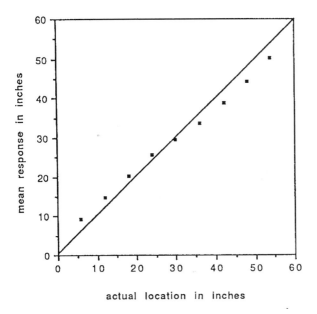

Figure 4.2
Mean response locations compared to correct response locations, from experiment 1 in Huttenlocher, Newcombe and Sandberg (1994).

cloth cushions covering the ground of a circular children's wading pool. Children searched with impressive accuracy, indicating ability to encode distance and direction. Because children were never required to search from a location other than the one where they sat to observe hiding, we cannot tell if coding was with respect to their own position or with respect to the circular borders of the pool. Nevertheless, the fact that infants at the end of the first year could remember and use distance information from any referent is an impressive finding.

Infants
The ability of children as young as 12 months to code distance and direction leads naturally to questions about the development of this ability in the first year of life. Given the flood of studies investigating the capacities of infants in the last decades, one might expect that a literature review might produce information on these abilities. We found, however, that although there are three potentially relevant kinds of studies concerning coding of extent in infancy, none of them precisely addresses infants' ability to code distance in continuous space. Hence we went on to gather some new data of our own.

A first kind of experiment that might be thought relevant to infant coding of location is the A-not-B situation examined in chapter 3. However, the fact that screens (or equivalent hiding devices such as cloths) are used in A-not-B studies makes findings difficult to interpret as bearing on coding of location in continuous space. In addition, as we have seen, A-not-B situations probably index ability to inhibit frequent and recent motor responses as much as they assess location coding with respect to a frame of reference.

A second kind of experiment involves examining infant looking time in situations in which objects are hidden behind screens. There are two problems with these studies. First, the findings from screened-location studies conflict with each other, showing success in location coding at ages ranging from 2 to 8 months. Baillargeon and Graber (1988) and Baillargeon, DeVos, and Graber (1989) found that 8-month-olds, but not 7-month-olds, could remember which of two screens an object had been hidden behind. By contrast, Wilcox, Nadel, and Rosser (1996) found that even 2-month-olds showed evidence of remembering which of two screens an object had been hidden behind. In a study in which a toy was hidden behind one of four screens arranged in a line, Wilcox, Rosser, and Nadel (1994) found that 6-month-olds showed evidence of coding location only in certain situations.[1] Second, however the studies using screened locations had turned out, their relevance for under-standing coding in continuous space would have been unclear. On the one hand, the presence of screens may make location coding easier than in situations involving a continuous expanse, since distinct screens may serve as spatial placeholders which facilitate infant coding. On the other hand, the presence of multiple identical screens could be distract-ing or confusing to infants.

A third kind of experiment involves infants' coding of the height of objects. In a series of studies, Baillargeon has found evidence that infants represent the shape and size of objects. For instance, Baillargeon (1987) showed that even 5.5-month-old infants can encode the height of objects, using a paradigm in which rabbits of varying heights moved back and forth behind a screen in which a window had been cut. The shorter rabbit ought to have remained invisible when passing behind the screen, being too short to show in the window, while the taller rabbit ought to appear in the window when passing behind the screen. The infants in this study looked longer when these expectations were

1. When objects appeared from behind an end screen, after having been hidden at one of the middle screens or at the opposite end, infants looked longer than in control conditions, although they did not show longer looking times when objects reappeared at a middle screen after having been hidden at one of the ends.

violated.[2] However, it is not known whether coding the physical dimensions of an object is equivalent to coding the distance between objects. Neuropsychological evidence suggests that coding various attributes of objects is a distinct ability from coding the location of those objects (Ungerleider and Mishkin 1982). While studies on these "what" and "where" systems have not specifically examined the question of which system codes physical dimensions of objects, there is some evidence from case studies that object dimensions such as height may be coded in a fashion distinct from coding interobject distance. Patients who can judge the length of a block, and even its aspect ratio, sometimes cannot judge the distance between two objects (see the review by Farah 1990).

Thus the existing literature does not allow for a definitive answer to the question of whether infants can code distance in continuous space. However, coding of distance has never been examined directly in continuous space. Therefore Newcombe et al. (in press) conducted a series of studies examining the abilities of 5-month-olds to keep track of the location of an object hidden in continuous space, namely, a 30-inch long sandbox. Our first infant sandbox study aimed simply to determine whether infants as young as 5 months could code location in continuous space in a very simple situation (see figure 4.3). An object was hidden in a certain position in a sandbox in full view of the infant and, after a short interval, dug out of that same position. After observing four events of this kind, some infants were shown the object being hidden, as usual but being dug out of a different position, 12 inches away. We found that infants react to such a demonstration with increased looking time, relative to a control group that saw the object emerge from the actual hiding location. The data from this study show that 5-month-olds are able to discriminate between a location where something has recently been happening and a location at which nothing has ever happened.

In a second study we sought to determine whether performance in the first study could be explained on the basis of categorical rather than

---

2. Bogartz, Shinskey, and Speaker (1997) have recently argued that studies of this type in general, and the study of infants comparing the height of the rabbit and the height of the window in particular, are susceptible to an alternative perceptual interpretation which does not require use of terms such as "expectation" or "reasoning." In their view, these studies examine perceptual routines and preferences rather than cognitive convictions. These suggestions are intriguing, and they may well provide a better account of these infant looking time experiments than the traditional one. (In chapter 5 we discuss the possibility of a perceptual to cognitive shift.) However, resolution of this argument is not crucial for our present purposes. The perceptual explanation still provides an interesting account of the early departure points for later spatial development.

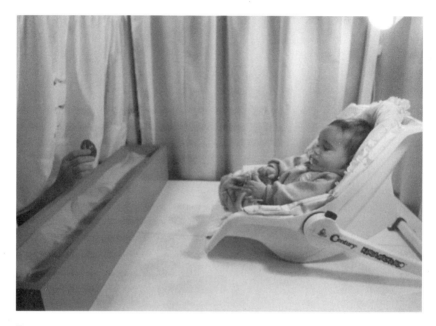

Figure 4.3
An infant looks at a decorated plastic egg about to be hidden in a sandbox. Procedure used in Newcombe, Huttenlocher, and Learmonth (in press).

fine-grained coding. The 12-inch gap between locations in the first study was fairly large, twice that of the 6-inch gap between locations studied in prior work with toddlers (Huttenlocher et al. 1994). Because the sandbox used was smaller than that in the toddler work, the gap may have loomed even larger. In particular, objects positioned toward the sides of the box were only 3 inches from the end, a position that may have enhanced the distinctiveness of the gap. In other words, the infants may have reacted to a rough-grained coding of distance as a middle versus end contrast rather than conducted a more fine-grained coding of continuous distance. However, we found that, even when the gap was reduced to 8 inches, so that objects hidden away from the midline were 7 inches from the end of the box and looked (to adults) much more clearly "in the box" than "at the end," infants reacted with increased looking when an object was removed from the wrong location.

Finding this kind of basic localization ability is intriguing. However, from these two studies, we do not know whether infants can remember a location even when various events have been occurring at various locations in the world. That is, to take the simplest case, if events have occurred at two locations recently, can young infants still discriminate between the locations? We therefore conducted a third study to determine whether infants can code the most recent hiding location of a single object, which they have seen repeatedly hidden, in alternation, in both of two locations. Five-month-old babies were shown an object hidden six times, three times in one location and three times in another 12 inches away. When the object was hidden a seventh time and then shown emerging from the other location, they reacted with increased looking, relative to a control group, even though that other location had been used as a hiding location three times already. That is, they remembered the most recent hiding location of the two possibilities.

Finding that 5-month-old infants can remember the most recent hiding location of an object is impressive. We do not know, however, from such data whether infants remember particular objects (i.e., with specified shapes, colors, or sound-making properties) as having particular locations, or simply remember where "something" was most recently hidden. To study whether they did or did not link particular objects to particular locations, in a fourth study, familiarization trials were used in which infants were shown one object (a pink-and-purple mesh ball containing a jingle bell) hidden at one location in a sandbox and a second object (a clear rectangular box containing multicolored beads) hidden at a second location, in alternation. One experimental group then saw one of the objects hidden in its usual location but dug out of the location previously used only to hide the other object. Reacting to this event, which can be called a *location violation*, would be expected given the results of the second study. A second experimental group, having seen an object hidden at its usual location, saw the object usually hidden at the other location dug out of the wrong location. If this group, which can be called an *object violation* group, were to show reactions in looking time as well as the location-violation group, one could conclude that infants coded the particular locations of particular objects. On the other hand, if the location-violation group showed reactions but the object-violation group did not, one could reach two conclusions: first, that infants code the most recent of two hiding locations, even when events have been occurring at both, and second, that infants do not code particular objects as having particular locations.

The data from the location-violation group showed, as expected, that the infants' looking times showed reactions when an object was revealed from a different location than that used for the prior hiding event. On the other hand, there was no indication from the object-violation group in this experiment that at 5 months infants were at all reactive to seeing an object emerge from a location at which they had recently seen an object very different in appearance hidden, suggesting that at this age infants do not remember particular objects as having particular locations.

This conclusion is supportive of the recent results of Xu and Carey (1996). Their studies found that, through the age of 10 months, babies seem to define an object more by its spatiotemporal attributes than by its static perceptual attributes, such as color and shape. Infants in their experiments watched as two objects appeared and disappeared, in alternation, on either side of a screen. When the screen went down to reveal the area behind it, the infants looked longer at displays containing those two objects than they did at displays consisting of just one of them, suggesting that the infants had interpreted the original display as a single object moving from side to side and changing its perceptual characteristics behind the screen. A similar result was obtained by Simon, Hespos, and Rochat (1995).

Our experiment addresses some criticisms advanced by Wilcox and Baillargeon (1997) of the Xu and Carey conclusion. Wilcox and Baillargeon suggested that the babies in Xu and Carey's experiment may have had difficulty mapping the oscillating appearance event onto the event in which the screen went down and drawing conclusions which could only be based on relating those two events. Wilcox and Baillargeon (1997) in fact reported that infants as young as 4 months looked longer at events in which an object (e.g., a ball) went behind a narrow occluder (i.e., not wide enough to hide two objects simultaneously) and then emerged as a different object (e.g., a box), as compared both to situations in which the same object emerged or in which there was an occluder wide enough to hide two objects. They argued that situations of the kind they studied call for event monitoring rather than event mapping, and that only event mapping poses a challenge for young infants.

However, if infants can use static perceptual attributes as well as spatiotemporal characteristics to individuate objects, then they should react with increased looking time to the sight of a perceptually different object being dug out of the sand at a location at which they have just seen a different object buried. We found, using the same familiarization events as seen by the location-violation group, that infants in the object-violation group did not react to a change in static perceptual attributes when the location remained constant. That is, even in monitoring a single event in progress, they showed the same kind of error observed by Xu and Carey, suggesting that, early in life, infants individuate objects based more on their spatiotemporal attributes than on their form and color.[3]

3. A control experiment showed that infants at this age could discriminate between the two objects used in the study. We also did another study to evaluate a suggestion made to us by Renee Baillargeon: that if the infants had a chance to see the two objects side by side initially, and thus notice that there are two objects and perhaps contrast their attributes, they would do better in object-violation conditions. However, even when the

Overall, the four experiments we conducted on 5-month-olds' coding of location in continuous space show both strengths and limitations in the way infants code objects and their location. A surprising strength in the infant repertoire is the ability to code location in continuous space. The limits of this ability need exploration, however. First, the experiments did not stress information-processing capacities greatly. The hiding interval was only 10 seconds, only one object had been hidden, and only one event had to be reacted to. Whether infants can code location in long-term memory, or when several things need to be coded at once, remains to be seen. Second, the experiments did not determine whether the referent used to code distance, at whatever grain, was the frame of the box or the self (or both). Experiments in which the box and/or the infant seat were moved laterally would be required to examine this issue.

Whatever the limits turn out to be on how infants code location in continuous space, that they can do so at all is an important departure point for their building of spatial competence. Indeed, their attention to spatiotemporal attributes is a dominant means for infants to construct their world, so dominant that they must apparently deal with what, from an adult point of view, is an odd limitation. That is, they apparently treat as individuated objects any bounded entities that exist along a traceable spatiotemporal track. Despite the fact that infants certainly are able to perceive the color and form of objects in the world, they seem not to use this information to individuate objects and track them over time. Data from Xu and Carey suggest that use of this information in addition to spatiotemporal information may emerge at about 10 months. It is interesting that infants younger than 10 months apparently can relate color and form features to ongoing spatial transformations of individual objects, making predictions about likely orientations of objects that disappear behind an occluder while twisting in space (Hespos and Rochat 1997; Rochat and Hespos 1996). Spatial movements of an object with respect to object coordinates may pose different issues for infants than tracking spatial locations of an object with respect to external landmarks.

Summary
There is evidence that infants code distance in continuous space by at least 5 months. While questions remain about the nature of the representation implied by infant looking-time experiments, attention to distance in continuous space seems, at minimum, likely to be a useful

two objects were shown together before the experiment began, infants showed no reaction to one object being hidden and a different object being removed from that location (i.e., the null effect was replicated).

perceptual precursor to the building of spatial coding of distance. By 12 months at least, children can code distance in continuous space in such a way as to support search for hidden objects.

Place Learning: Coding Distance Using Distal Landmarks

If infants can code distance in continuous space, as suggested by the research just reviewed, one wonders whether they use such codings in a system of place learning in which distances from various distal landmarks are used to fix location in space. The ability to code distance shown in our infant studies need not imply the existence of place learning, since distance in the contexts we use is likely defined with respect to the geometric frame of the sandbox, something that surrounds the search space within the visual field and remains continuously perceptually available. By contrast, place learning requires coding distance from distal landmarks, relational information usually requiring integrating across views and not supported by an immediate perceptual frame. Place learning also requires the coordination of more than one distance—at least two reference points are necessary to fix location in two-dimensional space, and two only suffice if they are differentiated so as to specify direction. In fact place learning is known to develop later in rat pups than cue learning, and it seems to depend on mature hippocampal circuitry, while other coding systems do not (see Nadel 1990 for a review and discussion). In this section we explore whether there is also a later emergence of this coding system than of other systems in humans.

There are several studies suggesting that place learning may emerge relatively late in human development (i.e., after the initial reweighting which we have argued takes up much of infancy), although the studies vary greatly in just what ages of emergence they define. The differences among the studies appear to depend, however, on whether the criterion for emergence is taken to be the age of achieving mature competence or the age at which an ability can first be detected.

One set of studies focuses on the age at which children achieve mature competence. These investigations have shown that children younger than about 5 years do not perform as well as older children and adults on keeping track of where they have been in an eight-arm radial maze (Aadland, Beatty, and Maki 1985); Foreman, Arber, and Savage 1984; Overman et al. 1996) and that children younger than about 7 years do not perform as well as older participants on two adaptations of the Morris water maze for use with human children: finding an object in a circular enclosure and on a very large open field (Overman et al. 1996).

A second set of studies focuses on the age at which one can first observe evidence of place learning being used at all. DeLoache and Brown (1983) placed a desired object in one of four identical containers, and placed the containers, in full view of the child, in various locations in the center of a living room. Thus, to find the object, children needed to code the location of its container using distance and direction from landmarks, such as furniture, in the room. Children as old as 26 months had difficulty finding objects in this case, at least compared to how well they did when toys were hidden under distinctive associated landmarks. Mangan et al. (1994), somewhat similarly, hid a toy in one of eight identical locations, arranged in a circle on a circular platform enclosed with curtains. Children were disoriented and asked to search for the object. Finding the correct location would require coding of distance and direction from distant landmarks visible above the edge of the curtain. Children younger than 24 months failed to find the object.

The conclusion that place learning cannot be seen at all until toward the end of the second year requires further examination, however. First, the tasks used by Mangan et al. and by DeLoache and Brown required discrimination among multiple identical hiding containers. Because children of this age are generally reliant on using associated contiguous beacons to locate objects (i.e., cue learning; see the discussion in DeLoache 1984), the presence of multiple identical such associates may be confusing; they may distract children from using distal landmarks they might otherwise use. Second, children in both studies moved a good deal about the space before being allowed to search. Hermer and Spelke (1994) have reported that when children less than 24 months are disoriented in space by being turned around several times, they fail to use even coincident landmarks to localize objects, a skill in use from the first year of life. If this is true, disorientation might certainly be expected to disrupt use of the more complex system of place learning as well, and place learning might be evident in children younger than 2 years if they were not disoriented before searching.

Newcombe et al. (1998) therefore examined the development of place learning in a continuous space and when children were not completely disoriented. Children in this study searched for objects in a continuous space (the 5-foot sandbox) in which they were not distracted by the presence of multiple identical hiding containers, which might have led to underestimation of place learning in prior studies. They looked after they went around the box to the opposite side, a movement that was sufficient to take the accuracy of their searches away from the ceiling-level performance we had found previously but not sufficient for disorientation. One group of children looked for objects in the sandbox

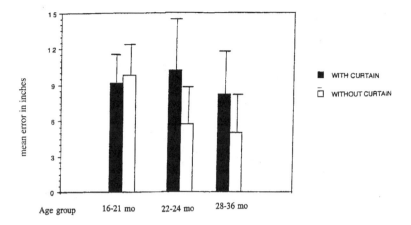

Figure 4.4
Mean errors for children of three ages with external landmarks hidden from view (dark bars) or visible (white bars), from Newcombe et al. (1998).

with visual access to external landmarks, while a second group did the task in an environment in which landmarks were hidden from view by a circular white curtain. The contrast between accuracy with and without visible external landmarks was crucial for investigation of the development of place learning. If there is a developmental transition, younger children would be expected to show the same degree of accuracy in search when they could see external landmarks as when they could not, whereas older children would be expected to show an improvement in accuracy when they could see the external world.

This study found evidence of developmental change between 16 and 36 months in the use of distal external landmarks (i.e., place learning). Specifically, the data indicated a remarkable contrast between the spatial coding seen in children of 21 months and younger and that seen in children 22 months and older (see figure 4.4). Children under 22 months did not refine the accuracy of their location coding using external landmarks when these are available, whereas older children did. This contrast was striking in size and apparent abruptness, and especially remarkable in that it was not affected by overall improvement in the basic task; younger and older children searched with equivalent accuracy when external landmarks are not visible. Apparently the results of DeLoache and Brown and of Mangan et al. were not due simply to children being confused when faced with multiple identical containers or having had their dead reckoning systems overwhelmed; in the present study children searched under the continuous surface of the sand and walked a relatively simple path around the sandbox. The fact that similar developmental findings emerge both when external landmarks are useful to refine spatial codings that otherwise lack precision (as in the present study) and when they are the only way to locate objects at all (as in DeLoache and Brown and Mangan et al.) suggests that the transition is a general one.

One possible explanation for the existence of an abrupt transition in the use of external landmarks, between the ages of 21 and 22 months, is that it depends on hippocampal maturation. This hypothesis is the one favored by Mangan et al. (1994). Support for it comes from evidence that similar transitions in performance by developing rats in the Morris water maze are due to such maturation, and that human hippocampal maturation continues until about this age (Kretschmann et al. 1986; Seress 1992). The fact that 21 months is also the age at which children become capable of performing a delayed-nonmatch-to-sample task (Overman 1990), a task that has also been linked to limbic diencephalic structures, including the hippocampus, adds some further support to this hypothesis. Diamond (1990; see also Diamond 1995) has argued that the late emergence of delayed-nonmatch-to-sample performance has more to do with the fact that success requires means-ends analysis (children must displace the novel object to obtain a reward) than with memory itself, and she has marshaled other data suggesting early maturation of hippocampus in human infants. However, recent data from animal studies show that DNMS performance may not depend on hippocampus but rather on perirhinal cortex, and that DNMS and place learning are doubly dissociable (Duva et al. 1997; Glenn and Mumby 1998; Kim et al. 1997). So the hippocampal maturation hypothesis could still be correct.

A change to use of external features in spatial coding might also depend on experience rather than, or in addition to, brain maturation. One possibility is that accumulating amounts of experience with upright locomotion, especially as it becomes increasingly less effortful, might be relevant to such a developmental change. It is difficult to see distal landmarks and a to-be-coded location simultaneously when crawling, and prewalking infants may simply not have the experience necessary to realize the value of such coding. While infants begin to walk at around a year, walking is initially tentative and effortful, and infants at the early stages of bipedal locomotion often look at their feet or at potential nearby supports more than they do at distal objects. The hypothesis that experience with skilled walking is necessary for place learning could be tested by comparing place learning in children who walk early and late.

Further study is needed to characterize the kind of external landmarks used and not used by children under the age of 21 months. Why do children of 16 months use the sandbox and apparently not use distal landmarks? One possibility to explain this contrast is to suggest that coding location with respect to an enclosing geometric shape is an autonomous kind of spatial coding (a "geometric module") (cf. Hermer and Spelke 1994 1996). But there are other possibilities as well. Perhaps children begin to use noncoincident landmarks that are fairly close to the to-be-coded location (e.g., the sides of the sandbox) and, with age, realize the value of using increasingly distant landmarks (e.g., objects several feet away, as are most trees or pieces of furniture). Or perhaps the need to code distance from two or more distal objects in order to get a fix on location is key to the developmental lag between distance coding and place learning. There are a variety of features that distinguish the sides of a box or the walls of a small room from discrete distal landmarks, and systematic investigation will be required to determine which are most relevant.

Summary
While many questions remain unanswered, the main lines of early spatial development are becoming clear. Not only do infants possess abilities to code location in terms of motor responses, contiguous cues, and simple forms of dead reckoning as discussed in chapter 3, they can also code distance in continuous space, at least over the short term and in simple situations. In fact they seem to focus rather fiercely on spatiotemporal information, weighting it more highly than color and form information in their definition of what an object is. Development occurs in the appearance of the ability to use distance codings with respect to distal landmarks and the consequent advent of a system of place learning in the second year of life.

Categorical Coding and Hierarchical Combination

In chapter 2 we described how, when people remember the location of an object, they may do so at different levels of detail. So, for instance, a person may remember that a friend's house is located about a third of the way up a block, that it is a few blocks away from the Famous Deli, and that it is in the Society Hill neighborhood. If asked to estimate the location in some way, the person will need to combine these memories to arrive at a judgment. If memory at a particular level is uncertain, then information at other levels will receive more weight in reaching an estimate. This hierarchical coding model, presented by Huttenlocher, Hedges, and Duncan (1991), describes the spatial location abilities of adults (Huttenlocher et al. 1991; Newcombe et al. 1999). In this section we explore its developmental origins. We begin by looking at its earliest manifestations, and then examine developmental change.

Developmental Origins of Hierarchical Spatial Coding

Toddlers
The sandbox studies reported by Huttenlocher et al. (1994) not only showed the early existence of fine-grained coding but also provided evidence for the early availability of a hierarchical coding system, involving categorical as well as fine-grained coding, and their combination across levels of coding. The evidence comes from the patterns of bias seen in the studies of toddlers' searching for objects hidden in a sandbox. As can be seen in the top panel of figure 4.5, children as young as 16 months systematically looked for objects, on the average, closer to the middle of the thin rectangular sandbox than they ought to have. The bias was somewhat less evident for points at the ends of the sandbox, as should be true if the children are able to code those locations as themselves distinctive. The other panels of figure 4.5 show that the bias toward the center remained in the other conditions examined in the sandbox studies: when the sandbox was surrounded by a circular white curtain which blocked children's view of landmarks, when children were moved laterally before search was allowed, and when children viewed the hiding and then initiated search from the end rather than the middle of the box.

The fact that the bias pattern continues to emerge in these conditions eliminates alternative explanations for the pattern that would be possible given only the original experiment. For one thing, a bias toward the center cannot just be due to children's being too lazy to walk toward the end of the box, since when they started search from the end (as in the bottom two panels) they actually had to walk past the correct location for the points near them in order to show this bias pattern. In addition a bias toward the center cannot be due to perceptual foreshortening, since it is seen even when children search from a different vantage point, from the point where they saw the original hiding.

Infants
Given the Huttenlocher et al. findings on hierarchical combination in 16-month-olds, one wonders whether even younger children might show evidence of such combination. Unfortunately, the only relevant research has examined the existence of categorical coding in infancy, without looking at combination across levels, and there are only a few studies even of categorical coding. What data we do have suggest that

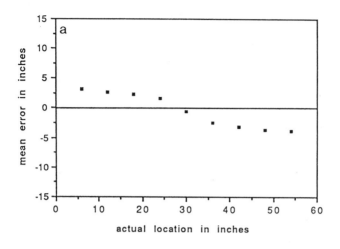

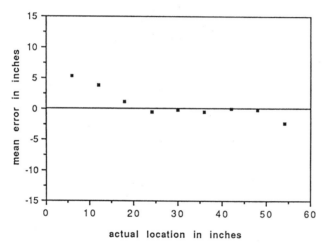

Figure 4.5
Bias patterns for toddler search in the sandbox, from Huttenlocher, Newcombe, and Sandberg (1994). Top left shows results when children viewed hiding and searches from a middle location in a room with visible landmarks. Bottom left shows data when a curtain enclosed the box; top right shows data when children viewed from the middle but moved to left or right before searching, and bottom right shows data when children both viewed and searched from a left or right position.

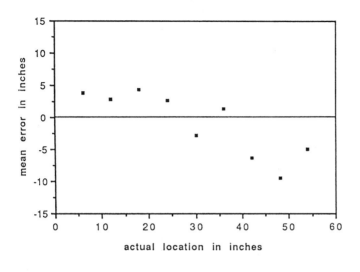

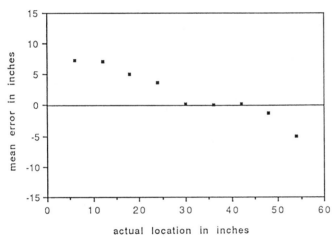

infants may group areas of space into categories, although perhaps only by 6 months or so.

Some early indications of the possibility that infants code space categorically come from studies investigating other topics, either object permanence or the relative position of two objects with respect to each other. In the object permanence literature, Baillargeon's (1986) study of infants' reactions to a toy car coming down a track when a block is located on or off the track can be considered evidence of categorical coding of objects as located "on" or "off" the track. In addition Antell and Caron (1985) studied infants' coding of relative position, showing that newborns have some idea of two shapes as being "above" and "below" each other. Newborn babies were shown an invariant arrangement of two shapes in a familiarization phase (e.g., a cross above a square), with the absolute position of the shapes on the stimulus card varying. They showed dishabituation when the position of the two shapes was reversed (e.g., square above cross). Similarly Behl-Chadha and Eimas (1995) showed that 3-month-olds coded the left-right relation of two individuated objects, using a design like that used by Antell and Caron.

However, none of these three studies addresses the idea of location in a region directly. What is needed is a study in which locations of objects vary, but, from an adult point of view, are all contained in a single spatial category or region. If infants code categorically, they should then dishabituate to an object shown in a location outside the category, even if the absolute distance away from prior locations is the same as that of a within-region object shown to a control group. Quinn (1994) conducted just such a study. He showed 3-month-olds a sequence of slides in which a dot appeared in one of four positions above (or, for another group, below) a bar. On a test trial, infants were shown two stimuli: one where a dot was shown above (or below) the bar in a new location, and one with a dot shown below (or above) the bar, in a location equally distant from the average of the familiarization trials as was the above-the-bar test dot. Visual preference was observed for the below-the-bar dot, indicating that that stimulus was perceived as more novel than the above-the-bar dot.

There are, however, limitations on infant categorical coding. One point is that referents, such as the bar in the Quinn experiment, seem to be required; Quinn (1994) found that 3-month-olds showed no visual preference for a dot in the opposite visual category to that shown in habituation trials if no horizontal line was present to demarcate the categories perceptually. A second limitation is that 3-month-olds do not form categorical groupings when the objects shown in habituation differ from each other, rather than all being dots, as in Quinn (1994).

Quinn et al. (1996) report that 6-month-olds, but not 3-month-olds, form categorical representations when varying shapes are used during habituation.

Summary
By 16 months children show the essential elements of a hierarchical coding system: not only fine-grained coding of extent, but also categorical coding of location, and combination across these levels of coding. In addition infants show categorical coding by at least 6 months as well as fine-grained coding at 5 months; hierarchical combination has yet to be assessed. It is possible that experience with how objects characteristically form distributions of locations with respect to spatial categories is necessary to drive the Bayesian process underlying hierarchical combination.

Developmental Changes in Hierarchical Spatial Coding

The studies we have just reviewed show the early availability of two essential components of hierarchical coding: coding of distance within an area and categorical coding. However, the categories used by young children may be restricted to ones that are very obviously perceptually available (e.g., the large wooden frame of the sandbox). Furthermore, in these studies, there was only one dimension to be coded. In this section we show that there is developmental change in two aspects of hierarchical coding: formation of categories that are not physically defined and hierarchical coding along two dimensions simultaneously.

Formation of Categories Not Physically Defined
The kind of hierarchical representation shown by the toddlers studied in Huttenlocher et al. (1994), while impressive, does not correspond to a mature representation. Adult patterns of bias are different from those seen with the toddlers in the sandbox. Errors for points in the left half of a rectangle move toward the center of the left half, and errors in the right half move toward the center of the right half. Thus errors move away from the center of the box, not toward it. In other words, adults code locations in rectangles not simply as "in the rectangle" but as being within one of two categories: the left or the right half. The halves are not perceptually given but rather mentally imposed categories.

We wondered when such use of nonperceptual categories would appear for coding in the sandbox. In a study conducted with children from 4 to 10 years (Huttenlocher et al. 1994, exp. 5), we found that children of ages 4 and 6 years still showed bias toward the center of the box, as had been seen in the toddlers. It was only at 10 years that we saw evidence that the sandbox had been organized into halves, with locations belonging to one of two categories (see figure 4.6).

One possible characterization of this finding is that children do not show any kind of mentally imposed subdivision of a perceptually given category until quite late in life. However, much earlier evidence of formation of mental subcategories within a bounded unit appears in a slightly different task. We asked children from 4 to 10 years to remember the location of dots within a small rectangle presented on a piece of paper. Clear evidence of division of the rectangle into halves for categorical coding was obtained in the 6-year-olds; even the 4-year-olds showed two-category coding as long as we scored their data so as to account for the effect of mirror-image errors (i.e., left-right confusions). Thus there is evidence that it is not mental subdivision of a perceptually bounded space per se that causes difficulties; it can be seen as early as 4 years in some tasks.

We do not yet have a good idea of what particular factors control the different developmental findings for the sandbox and the paper-and-pencil rectangle. The scale of the space is an obvious candidate, but there are additional differences between the situations, most obviously, the fact that the sandbox is an object, whereas the rectangle drawn on a piece of paper is not. An attempt to induce organization of a miniature sandbox into two halves by 4-year-olds by dividing the box perceptually (one-half blue, the other half red) did not create an adult kind of organization (Bud 1997). Instead, the obvious color difference seemed to reinforce the original categorization of the box as a single category or entity, perhaps by drawing attention to the prototypical center of that category, now salient as the red-blue boundary. Similarly adding lines or walls to define quadrants of a square did not help 7- or 9-year-olds to treat those regions as categories with central prototypes, although the devices did help 11-year-olds and adults to do so (Plumert and Hund 1999). In Plumert and Hund's study, the younger children seemed to treat the corners as prototypical locations, an interesting finding suggesting that perceptual salience (in this case of the corners) may continue to exert an important influence on spatial coding over an extended developmental period.

Before accepting the hypothesis that mental subdivision becomes increasingly more common and more differentiated with development, we should pause to consider a problem. The conclusion runs directly opposite to another well-known claim concerning the developmental course of subdivision. Kosslyn, Pick, and Fariello (1974) argued that children may have more subspaces than do adults, not fewer, perhaps because children define subspaces by linking locations they have actually walked between; this idea of action-based definition is quite

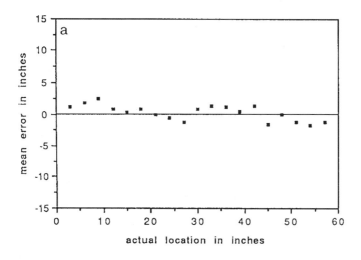

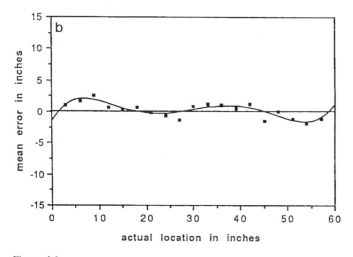

Figure 4.6
Bias patterns for 10-year-olds searching in the sandbox (top) and the same data as fit by
a polynomial function (bottom), from Huttenlocher, Newcombe, and Sandberg (1994).

Piagetian. Kosslyn et al. based their hypothesis on a study of the memory of 5-year-olds and adults for locations in a space divided into four quadrants by two kinds of barriers: opaque curtains and low fences. Finding evidence that the children showed distortions across both kinds of barriers (thus dividing the space into four subareas) while the adults showed distortion only across opaque barriers (thus dividing the space into halves), they suggested that, with development, space may become more integrated. However, their conclusion turned out to have more to do with the nature of their dependent variable than with location coding. In their experiment, children were asked about location by being asked to rank order each of nine locations for closeness to a tenth, referent location. Each of the ten locations was used in turn as a referent. Newcombe and Liben (1982) argued that children's difficulties in maintaining a stable referent accounted for the Kosslyn et al. data. Indeed, they replicated the finding using rank ordering, but showed that when children were asked simply to estimate interobject distances, no developmental interactions were seen. Thus Kosslyn et al.'s suggestion that children have more subspaces than adults turned out to be an epiphenomenon of young children's difficulties with the rank-ordering task (see also Newcombe 1997).

Categorizations Involving Two Dimensions
We have just seen that mental subdivision of perceptually given spaces is one line of development in the childhood years. Another line of development involves the fact that most spatial coding situations require the use of two dimensions, rather than just one. When we ask children to remember locations of points in the sandbox or in a rectangle presented on a piece of paper, we are examining their memory for location along only a single dimension. However, in only a slightly more complicated situation, namely in coding the location of a point in a circle, two dimensions are involved—as they are, of course, in almost all naturally occurring situations calling for spatial coding. Thus determining whether the competence children show in a basically one-dimensional coding situation extends to two dimensions is an important issue.

Adults remember both the distance of the point from the center of the circle (radial location) and the angular relation of the point to the radii of the circle (Huttenlocher et al. 1991). Furthermore adults remember location for both dimensions categorically as well as at a fine-grained level, in terms of the quadrants of the circle defined by vertical and horizontal axes. Because coding of points in a circle was already well understood for adults, Sandberg, Huttenlocher, and Newcombe (1996) investigated the developmental origins of hierarchical coding in two dimensions using this situation. We found that children from ages 5 to 10 years coded fine-grained location along both the radial and angular dimensions and that they also showed categorical coding along the radial dimension. However, categorical coding along the angular dimension was shown only by the 10-year-olds (see figure 4.7). Younger children did not bias their angular estimates toward the prototypical angle for each quadrant of the circle.

This lack of categorical coding for angular information is not due to the intrinsic difficulty of angular coding. When children were asked to remember only angular information, even the youngest children showed bias patterns indicating categorical encoding of angle (Sandberg et al. 1996). In addition, 7-year-olds code angle categorically when dots appear only on the circumference of a circle and coding of distance from the center is not necessary (Sandberg 1995). Thus it is specifically in a situation in which two dimensions must be encoded that younger children fail to show a constriction of their error variance along the angular dimension, which is shown by the age of 10 years.

The difficulty with two dimensions does not appear to involve children younger than 10 years having difficulty with imposing arbitrary frames of reference on objects. Sandberg (1995) showed that 7-year-olds fail to show typical patterns of categorical coding of angle, even when they are provided with horizontal and vertical axes which divide the circle into perceptually obvious quadrants. It may be that younger children simply have capacity limitations which prevent them from noticing the central tendency of two dimensions of variation at the same time.

If mental subdivision of spaces is not present until the age of 10 years in all but the simplest of coding situations, a fair question is, What causes such subdivision to emerge at all? A plausible, although speculative, answer is that it is driven by unconscious, but deeply rational, record-keeping on successes and failures in finding desired locations in a variety of circumstances. Thus, if success in locating objects is more common, or errors less severe, in situations in which subdivisions are smaller, a rational system will be increasingly likely to attempt such fine subdivision. Thus, one begins with a system in which subdivision is driven by perceptually (and perhaps functionally) available bases, such as division of houses into rooms, of classrooms into activity areas, and generalizes from success at location coding in those situations to situations with progressively less compelling bases for subdivision. In this respect, development in the spatial domain may be different from development in other domains, such as the acquisition of language. Much has been made of the fact that children do not receive feedback about the grammatical correctness of their utterances, and learn

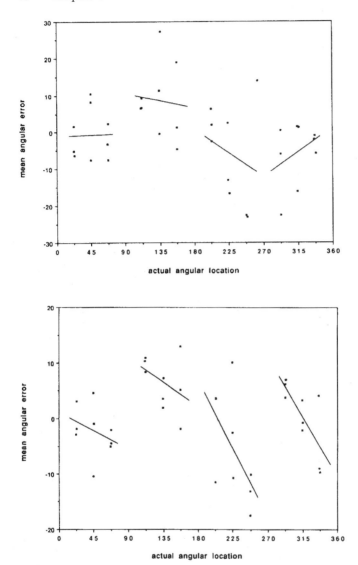

Figure 4.7
Mean angular errors plotted against true angular location with the within-quadrant regression lines for three age groups: 5-, 7-, and 9-year-olds in the top, middle, and bottom panels, respectively. Caption and figure from Sandberg, Huttenlocher, and Newcombe (1996).

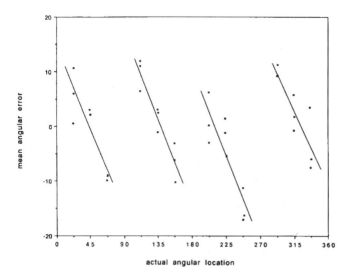

grammar nonetheless. For space, however, children probably do receive quite direct success and failure feedback, and they can modify their spatial coding systems utilizing such information.

How These Changes May Explain Developmental Improvement in Location Memory

The data on developmental change in hierarchical coding provide a means of reconciling theories that emphasize the early spatial competence of children with data showing that location coding undergoes lengthy developmental improvement. In the decade or so following publication of Siegel and White's (1975) chapter on the development of spatial representations of large-scale environments, rather consistent age-related improvement was seen in children's ability to judge distances, compare and plan routes, and construct and evaluate external representations of spatial configurations, including maps, diagrams, and drawings. Much of this improvement relates to either children's ability to operate on their location memory to make judgments, draw inferences, and so on (the subject of chapter 5), or their ability to deal with symbolic space (the subject of chapter 6). In this chapter, however, we concentrate on investigations in which children indicated their location memory in a direct or fairly direct way, for instance, by replacing objects in locations from which they had been moved or by pointing to unseen locations from particular vantage points.

One set of studies in this tradition concentrates on quantitative age-related improvement. An example of such a study is one reported by Anooshian and Young (1981). They studied children from the ages of 7 through 13 years, who had lived in the same new housing development for about the same period of time, two years on the average. The children were taken to four different vantage points around the development and asked to point a telescope to three well-known landmarks: the swimming pool slide at the recreation area, the gym set at the playground, and the door of a convenience store. Accuracy of pointing changed from about 27 degrees error for the youngest subjects to about 14 degrees for the oldest subjects. In addition consistency of pointing (i.e., the degree to which pointings from different locations converged on the same area) increased with age. Similar findings were reported by Anooshian and Nelson (1987) and Cousins, Siegel, and Maxwell (1983).

It is fairly clear how development of hierarchical coding can account for effects of this kind. As children become older, our studies suggest that they are likely to divide their neighborhood into smaller categorical areas. For instance, younger children might not divide the housing development into mental sections at all, or might, at best, mentally divide into a home area and the nonhome area. By the age of 13 years, they are likely to have organized the development into subsections, subsections that may revolve around exactly some of the targets used in the Anooshian and Young study: the recreation area, the playground area, the store area, as well as, perhaps, various friends' home areas, and the like. Because hierarchical combination reduces error, older children using mental organization of the development would be expected to have smaller pointing errors than younger children relying on fine-grained coding combined with category information from a very large and inclusive category. Similar arguments can be applied to the even larger pointing errors found in studies of preschoolers (e.g., Herman, Shiraki, and Miller 1985).

There are other studies showing age-related improvement in location coding which have emphasized qualitative rather than quantitative change. For instance, Herman and Siegel (1978) found marked age differences between kindergartners and fourth graders in their ability to replace target locations, apparently as a function of the availability of nearby landmarks. Kindergarten children performed quite well in a classroom setting with landmarks close by, but poorly in a gymnasium with only more distant landmarks available. Older children showed no variation as a function of setting. Herman and Siegel argued that the younger children's location coding was dependent on the presence of nearby landmarks, a theme they linked to Piaget's theory that children change from topological to Euclidean coding between the ages of 5 and 10 years. Similar conclusions were reached by Acredolo, Pick, and Olsen (1975).

However, the apparently qualitative change found in these studies turns out to be fundamentally quantitative in nature. Huttenlocher and Newcombe (1984, exp. 1) conducted a study based on the Herman and Siegel one. Objects were arrayed on a rug, which was either or not surrounded by landmarks. The dependent variable was the number of objects placed accurately, defined as within a foot of the correct location, as done by Herman and Siegel. Accuracy was indeed found to vary as a function of the availability of nearby landmarks, for 5-year-olds but not for older children, replicating the effects of Herman and Siegel. However, it was not true that the younger subjects who did not have nearby landmarks had no idea of the location of the objects. Memory for absolute location is not the same as memory for relative location (see also Naveh-Benjamin 1987). A close examination of the data showed that 5-year-olds who didn't have nearby landmarks showed somewhat larger errors, which were more likely to fall out of the arbitrary window for counting an object placement as "correct" (see figure 4.8). But when either a more stringent or a more relaxed criterion was employed, the interaction of nearby landmarks with age disappeared. Even more important, 5-year-olds showed high levels of configurational or relative accuracy, even with only distant landmarks available. Similar developmental results are found in Uttal (1994, 1996).

The age-related interactions reported by Herman and Siegel (1978), as well as by Acredolo, Pick, and Olsen (1975), appear likely to be due to changes in hierarchical coding, rather than a qualitative shift from topological to Euclidean coding. That is, in the absence of nearby landmarks, fairly large empty spaces of certain shapes may (or may not) be subjectively divided into subregions. Such subdivision is more likely in older subjects, and results in more accurate localization.

Summary
Hierarchical coding is evident by 16 months at least, but a mature system of hierarchical combination takes much longer to develop. It may not be until 10 years that children use spatial categories lacking perceptually obvious boundaries in large-scale spatial coding. Until this age they also lack the ability to perform hierarchical combination along two dimensions at once, apparently due to capacity limitations. The upshot of these weaknesses in children's spatial coding is that their accuracy on various tests of localization is a relatively rough one, and accuracy progressively improves during the elementary school years.

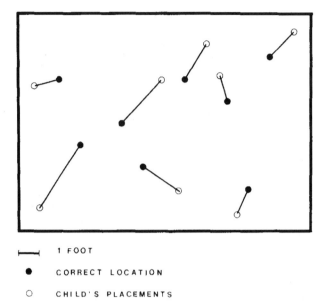

┣━━━┫   1 FOOT

●   CORRECT LOCATION

○   CHILD'S PLACEMENTS

Figure 4.8
Example of how someone can have good configurational knowledge but poor memory for absolute location. From Huttenlocher and Newcombe (1984).

Development of Dead Reckoning

Dead reckoning is as necessary as place learning for adaptive functioning, which could never be based on cue learning and response learning alone. The systems are similar in that they both incorporate metric information about direction and distance; dead reckoning does so with respect to self-reference, and place learning with respect to external landmarks. The dead reckoning system appears to be an autonomous mode of spatial reckoning driven by efference and proprioceptive information (Rieser 1989; Rieser, Guth, and Hill 1986).

As we saw in chapter 3, infants are able to deal with dead reckoning in simple situations. They can code amount of rotational movement about their own axis by at least 8 months (Landau and Spelke 1988; Lepecq and Lafaite 1989; Rieser and Heiman 1982; Tyler and McKenzie 1990), and code amount of movement along a straight line by at least 9 months (Landau and Spelke 1988). However, movements involving both translation and rotation, in which one both moves to a new location in space and changes one's heading, are by far more common in everyday life than translations alone or rotations alone, and such movements still pose difficulty for spatial localization at 9 months (Landau and Spelke 1988). As infants begin to crawl, their performance on problems in which they have to move to the other side of a display before responding improves, but exactly what is learned at first is subject to more than one interpretation.

Acredolo, Adams, and Goodwyn (1984) showed that rotation-and-translation problems, in which children move to the opposite side of a display, can be solved at 12 months only when children are allowed to track the correct location visually while moving. By 18 months, children succeeded on the task even when they did not look at the display. However, because Acredolo et al. used a problem in which a hidden object was placed in one of two clearly marked hiding places, the success of 18-month-olds could have been based on a simple reversal rule: "Following this kind of motion, choose the opposite location (defined in viewer-centered terms) than the one I did before." In this case, 18-month-olds's ability to code their own motion and adjust their location coding accordingly might not be truly well-developed.

In fact, Bremner, Knowles, and Andreasen (1994) showed that when use of a reversal rule was impossible because children were given four choices rather than just two, 18- and 24-month-olds made a very substantial number of errors. They performed at or near chance when the four locations were located close together and quite poorly, although above chance, when the locations were more widely separated. By 3 or 4 years, children performed much better. The Bremner et al. study leaves open the possibility that what 18-month-olds used to succeed in the Acredolo et al. study was actually a simple reversal rule, and that it is not until after 24 months that movement information is used in any precise metric way to update location.

In order to further assess the question of whether toddlers can use metric coding of the distance and direction of their own movement to adjust their estimates of the location of objects, Newcombe et al. (1998) observed children search for objects hidden in a sandbox after moving to the other side of the box. The use of a continuous space allowed children's accuracy to be metrically evaluated. The children did the task in an environment in which landmarks were screened from view by a white circular curtain. If children cannot use dead reckoning and rely on a simple alternation rule in solving two-choice tasks, they might be unable to locate objects in the sandbox at all after moving around it in a curtained environment. On the other hand, even the youngest children might use dead reckoning, differentiating among possible locations after moving. An intermediate possibility is that differentiation among locations could become more precise with age, indicating fine-tuning of the dead reckoning system.

The data showed quite clearly that, by 16 months, children have an above-chance capacity to perform dead reckoning following rotation and translation. Their searches for hidden objects, while significantly less precise than their searches when they remained stationary, were systematically related to the correct locations. It appears that Bremner et al.'s findings on 18-month-olds' difficulty with search for an object hidden under one of four covers do not reflect a complete inability by children of this age to relate spatial coding to self-produced movement which moves them to a new location and changes their heading. Rather, Bremner et al.'s data likely reflect the relative imprecision of the adjustments made by children at this age, an imprecision which would be especially problematic when possible locations were close together. Thus, between 12 months (when Acredolo et al. 1984 found that infants could only perform two-choice tasks when they could track the correct location with their eyes) and 16 months (the youngest age in this study), it seems that children do not simply learn alternation rules which allow them to perform simple two-choice tasks but begin to adjust their spatial coding in a metric fashion. That is, there is apparently considerable developmental change between 12 and 16 months.

Experience with upright walking during this time could be important to this development, as walking allows for a continuous view of the transformation in spatial location of objects as one moves. However, if walking experience is important to this transition, it has its effect swiftly and completely; there was no evidence that dead reckoning improved with age across the 16- to 36-month time span studied by Newcombe et al.

Although children in the age range we studied were uniformly above chance in locating objects after moving around the sandbox, there was a substantial dropoff in performance when they moved, as compared with performance when they remained stationary. This decrement did not change with age across the 16- to 36-month span studied, suggesting little change in the accuracy of dead reckoning in the time just after its first emergence. At least two changes may occur later in development. First, dead reckoning may not be used at first when visual cues are available. Rider and Rieser (1988) asked 2-year-olds to point back to their mother after they had been led away from her on a path involving forward motion and several right-angle turns (i.e., leaving a room, turning and walking down a hall, and turning and entering another room). The children were able to point correctly when they had traveled in the dark but not when external landmarks were visible; older children pointed correctly in either case. Success in the dark but not in the light may imply that at first, children use dead reckoning primarily when the absence of vision leads to a focus on efference and proprioception. Second, the fine calibration of the dead reckoning system may be a relatively late achievement. Rieser and Rider (1991) found increases in the accuracy with which people could locate objects after moving when blindfolded between the ages of 4 years and adult.

Summary
As motor abilities come on line due to maturation, the ability to code the amount of movement engaged in and use this information to adjust estimates of location follows behind quite closely. Nevertheless, there seems to be developmental change even after children are fully mobile, in the ability to ignore irrelevant or distracting visual landmark information and in the fine-grained accuracy of the coding of efference information.

Epilogue: Isn't Location Coding an Automatic Process?

We have argued that location coding, while beginning with fundamental elements in place early, also changes in important ways with age. However, some time ago, Hasher and Zacks (1979) argued that location is an automatically encoded attribute of items, using as one of their criteria lack of developmental change. They also suggested that a second criterion of automaticity was lack of improvement in spatial memory performance as a function of intention to remember. If true, their automaticity claim would sit very uneasily with everything we have just said about development. However, neither the claim of no developmental change in location coding nor the claim that incidental and intentional spatial learning are equivalent has held up well over time.

With respect to the first hypothesis, Park and James (1983) found improvement in location memory from first to fifth grade, Acredolo, Pick, and Olsen (1975) found greater accuracy across age, from preschool to 8 years, and age-related improvement has been reported through 13 years, as we just discussed (Anooshian and Nelson 1987; Anooshian and Young 1981; Cousins et al. 1983). In addition location memory appears to decline with aging (e.g., Light and Zelinski 1983; Naveh-Benjamin 1987; Pezdek 1983). With respect to the second hypothesis, although many people still believe that intention to remember has no effect on spatial memory performance, there likely is a small but reliable effect. Mandler, Seegmiller, and Day (1977), who are often cited as finding no effect of incidental versus intentional instructions on location memory, actually found a reliable effect. Better performance when people knew they would be tested on spatial location than when they saw it incidentally has also been found by many other investigators (e.g., Acredolo et al. 1975; Lansdale 1998; Light and Zelinski 1983; Naveh-Benjamin 1987; Park and James 1983). Thus location coding apparently improves both with age and with intention to learn, violating two of Hasher and Zacks's criteria for automaticity. In addition Naveh-Benjamin (1987) found that memory for spatial location improves with practice and is reduced by competing task demands.

Giving up on automaticity in the strict sense should not, however, mean forgetting some remarkable facts about spatial memory. First, even though younger children and the elderly do more poorly than older children and young adults in these studies, no group performs at extremely low levels. Second, incidental learning of spatial location is far above chance, and quite impressive given that spatial information is often coded in situations where its usefulness is not at all obvious (e.g., position of a picture on a page). Such findings show that location coding is a fundamental aspect of living in the world, and therefore one that may have an essential core of automatic procedures. As Logan (1998) points out, if visual attention is mediated by location, then paying attention to something may have as a prerequisite attending to its location. However, this automatic attentional core may be refined and supplemented when location is known to be important. In this case conscious strategies can be invoked, such as verbal coding and rehearsal of location, which would be expected to significantly increase memory for location. Such strategies are known to become more common during the early school years. In line with this approach, Lansdale (1998) presents evidence suggesting that the location memory shown in incidental conditions is inexact, a memory for a general region rather than a precise localization. He shows that it is only in intentional conditions that people show above-chance levels of exact recall for location.

Summary and Conclusion

We have now completed a sketch of what we think is now understood about the nature of location coding and its development. We believe that infants are likely born with the ability to code location several ways: in terms of motor responses, in terms of coincident cues, in terms of data from efference, and in terms of distance in continuous space (at least when a surrounding frame is available as a referent). Such claims are not radically nativist, however, in that they are tempered with a clear recognition of the ways in which these beginning abilities are limited.

During the first year of life, there are several vital lines of development. First, reweighting of the existing systems occurs so that they interface effectively. This reweighting is based on feedback from relevant experience with the physical world, feedback that is in turn partially paced by motor maturation and alters the nature of the situations the child encounters. Second, infants also acquire the ability to group together different objects to form a spatial category such as "above" or "below." Third, infants need to learn to define objects as adults do: to link the experience of particular static perceptual characteristics (e.g., shape, size, and color) with particular spatiotemporal pathways.

During the second year of life, there is additional developmental change, notably the advent of using distance information from stable external landmarks to define location (place learning) at about 21 months. The relative lateness of this accomplishment may depend on maturation of hippocampal areas supporting relational learning and/or on experiential factors. While ceiling-level performance on place-learning tasks is not achieved until the elementary school years, the emergence of the ability to use distal landmarks at all is a notable accomplishment.

Subsequent to the first two years, there is further development in the dead reckoning system, consisting in fine-tuning and calibration, as well as a suppression of visual information when it is misleading. There is also substantial developmental change, over an extended developmental period, in the kind and size of categories used in hierarchical spatial coding. These changes likely account for the age-related changes in the accuracy of location coding seen in some classic studies of the development of spatial representation.

Our account of the development of location coding differs in many respects from that of Piaget. Piaget was a strong advocate of the position that developmental changes in spatial representation are qualitative in nature and occur late in development, at the age of 9 or 10 years. He argued that prior to this time, children's coding was topological, that is, only encoding relations of enclosure, touching, and proximity. At the age of 9 or 10, he believed that children construct two new spatial coding systems: the projective system, encoding the information obtainable from a variety of perspectives, and the Euclidean system, encoding spatial information in terms of extent along invariant horizontal and vertical axes. We do not find this typology of spatial coding useful, so we have adopted a different one, which we think makes more contact with the current literature on adult cognition, animal learning, and neurophysiology. In addition we see much evidence that spatial coding is more sophisticated at earlier ages than Piaget believed, specifically, not only topological but also incorporating metric information.

However, these divergences from Piaget should not be allowed to obscure certain similarities. In particular, we see good evidence of changes in spatial coding that continue through the first ten years of life and that probably underlie some of the phenomena Piaget identified. At a more abstract level the account of development we advocate is interactionist, rather than focusing primarily on either native endowment or imprinting from the environment. We have attempted to show how initial departure points create an organism that is able to take advantage of the kind of information available in the environment and create progressively more adaptive methods of fixing spatial location.

It is somewhat premature, however, to discuss divergences and similarities of our position and that of Piaget, since thus far we have concentrated on spatial coding. Hence we have not addressed what Piaget most cared about: representation that would support reflection. In using terms such as spatial representation, Piaget drew a sharp distinction between representation at a level that would support action in the world (i.e., what we have been calling coding) and representation that would support reflection about the world (i.e., spatial thought). We turn next to the topic of spatial thought and spatial problem solving.

Chapter 5
Development of Spatial Thought

So far in this book we have considered how people remember the location of objects and how this ability develops. When experimenters ask people to find a hidden object, or to put an object back where it was before (or when people decide themselves to look for their keys or to put away their hat), we see people's knowledge of spatial location revealed in a relatively direct way. That is, people perceive and encode an environment of objects and referents, and they access these representations as necessary in order to establish location. While location estimations may be uncertain, and may involve unconscious combination of information across levels of coding or frames of reference, finding things does not require extensive operations on stored spatial information. For instance, there is no need to draw inferences from one's knowledge or to map estimations onto verbal descriptions.

However, people also need to think about, and act on, spatial location information. They may need to imagine how scenes would look from other vantage points, to compare distances, to plan routes around environments, or to establish sensible areas of search for lost objects. In fact, almost any interaction involving large-scale space requires inference, since a person looking at a large area from a single vantage point cannot take in the whole scene. Multiple views are required to learn about a large-scale space, and hence people must combine spatial information acquired on separate occasions to learn about such a space.

In analyzing development, there has been controversy regarding how to interpret children's difficulties on spatial-cognitive tasks. Piaget used children's problems with tasks such as spatial perspective taking to argue that children had a fundamentally different kind of spatial representation than do adults. In doing so, he did not distinguish between the possibility that children found the tasks hard because of the nature of the information they had coded about the world, and the possibility that they found the tasks hard due to differences in cognitive competencies such as logical thought and inference ability. Subsequent research, to be reviewed in this chapter, has shown that children's performance in perspective taking and in other spatial judgment tasks can be explained without assuming fundamentally different spatial representation.

Findings of early success on cognitive tasks are often taken to support some version of nativism (e.g., Landau, Gleitman, and Spelke 1981; Landau, Spelke, and Gleitman 1984). However, to argue for innate Cartesian competence on the basis of such data would be unjustified. There is some reason to suspect that children younger than 2 years or so would be unlikely to accomplish any of these tasks, and that a fundamental change from a short-lived perceptual spatial coding of location to a more durable representational coding may occur toward the end of the second year of life. In addition there are methodological and conceptual flaws with the key papers usually taken to support innateness in the spatial domain.

In this chapter we begin by discussing the basis for our suspicion of an important qualitative change at the end of the second year and offer a critique of the studies often taken as support for a nativist position. We then deal with the development of the ability to perform four spatial-cognitive tasks. Two of these tasks were central to Piaget's claims about spatial representation: children's ability to take the spatial perspective of others, and children's judgments of distance and length. The other two tasks have been investigated more recently, and from a wider variety of theoretical viewpoints: how children find their way around environments, and how children establish logical search patterns. We conclude by discussing the central theme of the chapter, namely that although the specifics of Piaget's theory of spatial development were wrong, the default need not be to nativism. Instead, we can construct a new interactionist account of how spatial thought evolves during development.

Does Spatial Coding Constitute Spatial Representation?

Piaget drew our attention to many changes in children's behavior occurring between the ages of 1 and 2 years, arguing that they indicated the beginning of a capacity for symbolic thought that had been lacking at the beginning of life. These changes include the beginnings of linguistic reference, symbolic play, and the ability to solve invisible displacement tasks. More recently, some investigators have suggested that representation is available earlier. For instance, experimenters have used observations of infant looking time to argue that infants have representations of hidden objects and expectations about their size and solidity (e.g., Baillargeon, Spelke, and Wasserman 1985), and they have seen evidence of representation in deferred imitation observed in children as young as 9 to 11 months (Meltzoff 1988; Mandler and McDonough 1995). These claims of very early representation are not uncontested, however. A variety of critics have pointed out that observations of looking time and deferred imitation may be explained on perceptual grounds (Bogartz, Shinsky, and Speaker 1997; Haith and Benson 1998; Müller and Overton, 1998; Thelen and Smith 1994) and have suggested that representations may be a graded, rather than an all-or-none, phenomenon (Munakata et al. 1997). A graded approach allows one to accept that infants as young as 3 months have declarative memory (e.g., Pascalis et al. 1998) without necessarily granting that such memories can be manipulated in thought.

If one rejects an all-or-none approach to the controversy over infant representation, a natural question to ask is whether there is evidence that infants' memories for spatial information can be used in various kinds of thinking. At present, there is little positive evidence of such a capability until after 12 months of age. Even in a simple "perspective-taking" situation, in which children walk to the opposite side of an array to locate an object, success is not seen until between 12 and 16 months (Newcombe et al. 1998)—and this is a situation where children's physical actions support the drawing of conclusions about location, so that it may not qualify as inference at all. Between 15 and 18 months, infants begin to be able to reason about the movements of hidden objects, much as Piaget initially suggested (Haake and Somerville 1985). In addition other findings indicate an ability to compute shortest routes, even when the goal is not visible, emerging at about this time (Pick and Rosengren 1991; Rieser and Heiman 1982). At about 21 months there is apparently a shift, perhaps an abrupt one, from an inability to use distal landmarks to locate objects to their use (Newcombe et al. 1998). Since place learning is the distinctive behavioral signature of spatial coding using the hippocampus, it is at least possible that one of the most important types of spatial coding does not appear until this time, and hence that whatever spatial thought precedes the appearance of such coding is distinctively different from the kind of spatial thought requiring such information.

It is also important to remember that all of the experiments used to examine spatial location coding in infants and toddlers have used relatively short delay periods. In infant looking time experiments, the delay periods vary from a few seconds to several tens of seconds, but to our knowledge they never exceed a minute. Hence it is possible that the coding observed in these studies is supported at the neural level by continuous firing of cells in the dorsal stream of visual processing (Gnadt and Andersen 1992; also see Haith and Benson 1998 for discussion) rather than by long-term traces. In experiments with toddlers looking for objects in the sandbox, the delays are more variable, because the procedure is somewhat paced by the child, but the mean delays are 10 seconds or less, and even long delays do not usually exceed a minute. While there have been anecdotal reports of long-term spatial memory in infants toward the end of the first year of life (Huttenlocher 1974), most of these observations involve situations in which a desired object had been repeatedly observed in a specific location marked by a prominent visual cue. Thus these observations may simply reflect an early ability to learn a paired associate with sufficient practice.

In summary, there has really been very little investigation of whether what infants know about space is a short-term perceptual encoding of information or whether they are also capable of forming longer-term representations that could be used as input to symbolic manipulation (i.e., thought). The transitions seen so far in the second year of life may indicate that spatial thought indeed has its origins at that time. Earlier development may lay the groundwork for the emergence of this capacity.

But Isn't Spatial Inference Innate?

There are extremely well-known claims that spatial coding has an innate basis (e.g., Spelke and Newport 1998). Such an argument rests on two prominent empirical foundations, namely the claim that sophisticated spatial development appears early and without significant environmental input (Landau, Gleitman, and Spelke 1981; Landau, Spelke, and Gleitman 1984) and the claim that early spatial coding contains an impenetrable geometric module (Hermer and Spelke 1994, 1996). Here, we examine these claims.

Early Appearance
Landau et al. (1981, 1984) report research conducted with a blind child of 2.5 years, called Kelli. Kelli was tested in a room containing four objects, one at the midpoint of each of the four walls. She was taught the paths between some pairs of objects, but not all. (See figure 5.1.) Given this acquisition experience, she was reported to be able to infer the paths to be taken to link novel pairs of objects. Such a performance, Landau et al. argued, indicated a Cartesian understanding of space because only an encoding of distances and angles for the known pairs would allow for the spatial inferences Kelli made. Furthermore they argued that it was striking and surprising that Kelli was able to code

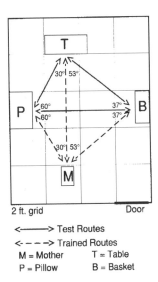

Figure 5.1
Layout for task used with Kelli by Landau, Spelke, and Gleitman (1984).
From Gallistel (1990).

such information (and use it as a basis for inference) in the absence of vision, since her blindness prevented her from perceiving any two landmarks in the room simultaneously.

These observations have been widely accepted in the literature as an argument for innateness. For instance, Gallistel (1990) cites this work as showing the spatial competence of young children, Geary (1995) uses the work to justify the conclusion that certain aspects of sex differences in mathematics performance are built on innate foundations, and Spelke and Newport (1998) regard the research as a foundation of a nativist approach to spatial understanding. However, there are many problems both with the data and with the conclusions drawn from them.

From a theoretical point of view, the crux of the argument from Kelli's performance is that Kelli's ability to code spatial information and perform spatial inference, in the absence of visual input concerning the spatial layout of the environment, indicates by default the presence of innate notions of spatial metrics and Cartesian relationships. But Kelli does not lack senses other than the visual. If she truly has the abilities that Landau et al. described, this fact may show simply that auditory, olfactory, tactile and proprioceptive input are sufficient to allow, over 2 or 3 years, for the development of spatial coding and inference abilities. If such abilities can develop in the absence of the input clearly best suited to foster them (see Thinus-Blanc and Gaunet 1997), this is a remarkable phenomenon, but it is hard to see how it is a proof of innateness.

However, arguing about how to interpret Kelli's performance may be somewhat premature, since there are empirical problems with the Landau et al. findings. One issue is that the data from Kelli are not in good agreement with other research on the spatial abilities of blind individuals. Congenitally blind individuals generally have difficulties with spatial tasks, including quite basic spatial matters such as encoding distance and the angle subtended by objects, which varies as a function of distance (e.g., Arditi, Holtzman, and Kosslyn 1988). They tend to represent routes in large-scale space as a progression of locations, rather than having an overall path representation, whose formation may be dependent on being able to take in several distant locations in a single glance (Iverson and Goldin-Meadow 1997). The degree of spatial impairment experienced by blind people is usually related to whether or not individuals had early vision (see Warren 1994 or Millar 1994 for reviews). Many comparisons of congenitally blind with sighted and visually impaired individuals indicate that some measure of visual experience, especially early in life, is necessary for normal ability to gain spatial knowledge about the environment and construct accurate Euclidean relations (e.g., Bigelow 1996; Rieser et al. 1992), although there may be some exceptions (see the review by Thinus-Blanc and Gaunet 1997). So Kelli, blind since birth, would be exceptionally spatially talented were her performance truly comparable to that of sighted children.

But was her performance comparable? Methodological critiques by Liben (1988) and Millar (1988, 1994) point out that the Landau et al. procedure allowed for a good deal of inadvertent cuing from auditory and olfactory cues, as well as self-correction by Kelli once she realized she had actually gone to the wrong goal. Linked to these methodological problems was the fact that the primary measure of Kelli's ability to make spatial inferences was her position when she stopped searching. Of course, prior to stopping, Kelli may have engaged in a good deal of exploration, self-correction, and on-line hypothesis-testing. Indeed, an inspection of Kelli's published paths shows that they were far from direct (see figure 5.2). Although some of the methodological flaws were corrected in later work with Kelli (as pointed out by Spelke and Newport 1998), by the time these controls were used, Kelli was much older, and may have learned to make the simple spatial inferences asked of her through her extensive experience with the situation and the task (see Millar 1994). A recent attempt to replicate the Landau et al. result has, in line with these critiques, shown rather different results.

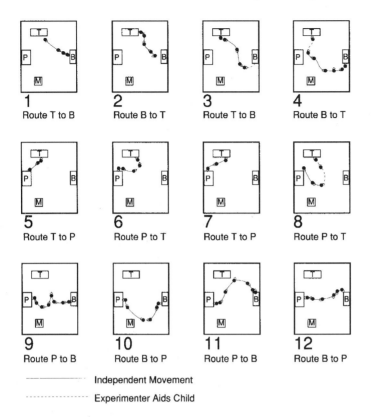

1 Route T to B   2 Route B to T   3 Route T to B   4 Route B to T

5 Route T to P   6 Route P to T   7 Route T to P   8 Route P to T

9 Route P to B   10 Route B to P   11 Route P to B   12 Route B to P

———————— Independent Movement

- - - - - - - - Experimenter Aids Child

Figure 5.2
Paths taken by Kelli in Landau et al. (1984) study. Kelli's positions are shown by solid circles, her directions by arrows. From Gallistel (1990).

Even using the final-position measure, Morrongiello, Timney, and Humphrey (1995) found that the blind children they studied were worse than age-matched sighted children in terms of accuracy at final position.[1]

While the data from Morongiello et al. undermine the argument made by Landau et al. that blind children were equivalent to sighted children in a spatial inference task, it is noteworthy that both the study with Kelli and the data gathered by Morongiello et al. suggest that blind children and blindfolded sighted children have a rough knowl-

1. Sighted children younger than 4 years, asked to wear blindfolds by Morongiello et al., were unwilling to perform the inference task at all. Differences between blind and sighted children were thus found in children considerably older than Kelli was, which further undermines the argument that spatial inference is early and easy.

edge of location gained from inference, in the absence of vision. When children (both blind children and sighted but blindfolded children) are asked to point to a location that must be inferred, they do better than chance, and, in Morongiello et al.'s data as well as Landau et al.'s, do not differ from each other. This knowledge is not as precise as implied in the Landau et al. discussions, but such findings are also at variance with claims by Piaget that early spatial coding is completely nonmetric or nonconfigural. In addition the existing data suggest that although visual information supports the most efficient spatial inference, such inference is ultimately possible even in its absence.

In summary, the existing data on spatial inference in blind children suggest that development in this area proceeds more slowly than in sighted children, although constructing spatial relations on the basis of nonvisual experience is eventually possible. The total picture is one of abilities increasing over time, rather than early mature ability in the absence of relevant input.

Modularity
The fact that basic spatial principles are necessary aspects of the physical world makes it tempting to conclude that use of such principles might not only be innate but might constitute a cognitive module. Gallistel (1990) advanced a modular approach to spatial coding, relying heavily on a series of investigations of spatial coding in the rat conducted by Ken Cheng (1986). In Cheng's work, rats learned to dig for food in one of the corners of a rectangular enclosure (see figure 5.3). When they were disoriented, they were equally likely to dig in the two corners that were both defined geometrically by the relation of the long and the short sides. Remarkably the rats showed this pattern of search even when strong cues existed as to which of the two corners was correct, for instance, when one of the sides of the enclosure was painted a bright color, or when a distinctive odor marked the correct place. Gallistel argued that this imperviousness to cue learning showed the operation of an autonomous geometric module.

Hermer and Spelke (1994, 1996), based on Gallistel (1990), have argued for a modular view of early spatial coding in humans, suggesting that this modularity is cognitively penetrated only slowly in the course of development. Basically their studies with toddlers paralleled those conducted by Cheng with rats. They found, as had Cheng, that after disorientation, toddlers searched in areas defined geometrically, ignoring potentially helpful environmental cues. These cues were, however, used by human adults. Wang, Hermer, and Spelke (1999) have replicated these findings even for children who had extensive experience in a room containing three white walls and one colored wall; the

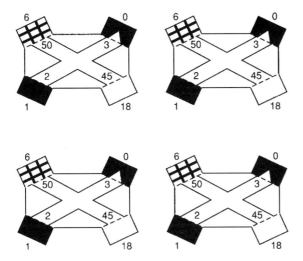

Figure 5.3
Experimental setup used by Cheng (1986), together with data for four individual rats. Each rat clearly discriminates between the two diagonals, but no rat uses the distinctive markings in each corner to discriminate between the two ends of the correct diagonal. From Gallistel (1990).

children did not use the colored wall to help them decide which corner a toy had been hidden in, after they had been disoriented.

The empirical phenomena reported by Hermer, Spelke, and Wang are interesting and important. In particular, it is fascinating to find that very young children code the length of walls well enough to use their relative length in combination with the handedness of the length ratio to find objects (i.e., they look for a hidden object in a way based on whether the shorter wall is to the left or the right of the longer wall). But do the findings support modularity? Here the critical claim is the lack of use of landmark information. But this finding is quite odd. When the place learning and dead reckoning systems conflict, the externally referenced system generally takes precedence, resetting both patterns of firing in hippocampal cells and actual spatial behavior (Etienne et al. 1985; Goodridge and Taube 1995).

An alternative is that the lack of effect of landmarks found in the Hermer-Spelke studies is due to the fact that the "landmarks" did not really qualify as landmarks, in that they were not stable. The curtain that created the "blue wall" was played with by the children in some experiments. In other studies, the wall was not always present in the room, since the manipulation was a within-subjects one. The toys used as landmarks in yet other experiments were small, movable, and, indeed, moved in the course of the study. Learmonth, Newcombe, and Huttenlocher (under review) have recently found that when truly permanent-looking landmarks (a door and a recessed bookcase) define diagonally opposite corners, toddlers are able to use this information to find toys even after effective disorientation. They are also able to use a single landmark (a recessed bookcase). And, especially challenging to the Hermer-Spelke-Wang position, children in our laboratory do seem to use a colored wall to reorient after disorientation.

Whether or not stability of landmarks is the key to their use has not yet been directly investigated. Wang et al.'s (1999) report that extensive experience with a colored wall did not lead children to use it in a search task following disorientation is a challenge to this surmise. However, these experiments used very few children (four in one study and four in another), so the matter clearly needs further attention. Another way to explain the contrasting results is based on noting that the scale of the spaces in the studies was different. The Hermer-Spelke research with a rectangular room used a 6 by 4 foot space, a crowded room when occupied by two or three people, and a very small area compared to ours (12 by 8 feet). It might be that landmark use is only activated when areas are at a scale affording some analogue of real-world navigation.

The idea that a geometric module is a ubiquitous feature of mammalian functioning has suffered another recent blow in the findings of Gouteux, Thinus-Blanc, and Vauclair (1999). These investigators report that rhesus monkeys were able to use a colored wall to find a hidden food reward after disorientation, both when the wall was directly associated with the food and when the wall constituted an indirect landmark. The same seems to be true of cotton top tamarins (Gouteux, personal communication, May 17, 1999).

Development of Spatial Perspective-Taking

One of the most famous of Piaget's experiments concerned whether young children can take the visual perspective of other observers. As almost every introductory child development textbook explains, Piaget showed children of various ages a display consisting of three distinctive mountains (shown in figure 5.4) and asked them to indicate what an observer located at another vantage point would see of the mountains. He sometimes asked them to do this by choosing among pictures or models of alternative views, sometimes by building a model themselves. Irrespective of these procedural variations, Piaget found a consistent developmental picture: children younger than the age of 9 or 10 years did not seem to be aware of what the other observer would see. Indeed, younger children were sometimes apparently not aware that

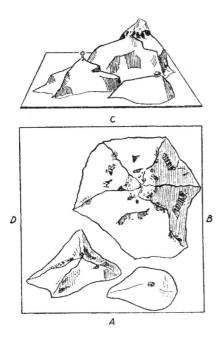

Figure 5.4
Setup for Three Mountains task, from above and from the side. From Piaget and Inhelder
(1948/1967).

the other observer saw anything at all different from what they saw. They indicated instead that the observer's view would be the same as their own view. Such errors, called "egocentric" errors, were said by Piaget to be an indication of the general egocentrism of early childhood thought. Furthermore Piaget argued that the change in children's perspective-taking performance showed a fundamental change in spatial representation, from an early "topological" form of spatial representation, in which only relations such as "closed or open" and "touching or separated" were encoded, to "projective" spatial representation, in which the orders of objects encountered along specific lines of projection were encoded.

A substantial literature, consisting of hundreds of empirical articles, has grown up around the perspective-taking task and the associated claims Piaget made about it (see Newcombe 1989 for a complete review). For some time now, it has been recognized that several aspects of Piaget's analysis were wanting. First, the claim that there is a general construct of egocentrism has received little empirical support (see the review by Ford 1979). Visual perspective-taking tasks are not strikingly correlated with communicative or affective role-taking tasks, such as performance in describing unfamiliar visual forms to someone else who can't see them or performance in guessing the emotions others would feel in certain specified circumstances.

Second, the claim that children show egocentrism in the visual perspective-taking task, even considered alone, has also been unsupported by empirical evidence. For instance, to offer three well-known examples from among the many studies making this point, Flavell, Omanson, and Latham (1978) showed that young children did understand that other observers saw something different (they just weren't sure what), Lempers, Flavell, and Flavell (1977) showed that children as young as 2 years displayed a drawing so that other observers could see it (i.e., held it up away from themselves), and Hughes and Donaldson (1979) showed that preschoolers looking at a stage containing an arrangement of walls and open spaces, as well as two dolls described as a thief and a policeman, could predict when the policeman could see the thief and when he couldn't. The ability of very young children to overcome visual egocentrism may be based on a line-of-sight heuristic (Yaniv and Shatz 1990). That is, children appreciate that observers can see whatever objects or parts of objects for which it is possible to draw a straight line between the observer's eyes and the object, without that line crossing an opaque obstruction. This heuristic allows for success on many tasks; for instance, in Hughes and Donaldson's study, children could trace a mental line between the two dolls and see if it crossed a wall or not in order to determine if the policeman could see the thief.

Third, the idea that children have difficulty computing other observer's viewpoints is also questionable. While it would be possible for young children to be nonegocentric (i.e., to know that other observers see something different from what they see) and yet lacking in the spatial-representational base or the inferential ability needed to work out exactly what the other observers do see, this does not appear to be the case. The key difficulty in the perspective-taking task seems to be the fact that answering the question by picking from a set of alternative views, or by making a model, entails dealing with conflict between two frames of reference.

Consider the classic perspective-taking situation, as shown in simplified form in figure 5.5. A subject sits facing a table, with a ball positioned on the left. The subject, the table, and the ball are all, in turn, situated within a room that contains, say, a desk and a cabinet, and also has a door and a window. Coding the location of objects such as the ball can be done with respect to the subject or with respect to the circular tabletop. But the most natural and powerful method for locating the ball, and one in place since the end of the second year of life,

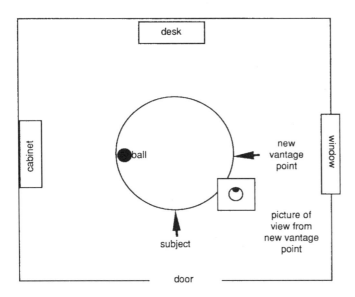

Figure 5.5
Example of room, array and picture in a picture-selection task. From Newcombe and Huttenlocher (1992).

is to relate it the external frame of reference, namely the desk, cabinet, door, and window. Now suppose that the subject is asked to indicate what an observer at a new vantage point would see, by pointing to a picture of that view. (Figure 5.5 shows only the correct alternative, for the sake of simplicity, but actually at least three incorrect views would usually also be presented.) The correct picture is presented, naturally enough, so that it faces the subject. This positioning entails that the picture shows the ball in the incorrect position with respect to the existing frame of reference (with the wrong relations to the landmarks of the desk, cabinet, etc.). Picking this picture as the view from a new vantage point thus requires resolution of the conflict between the perceptually present frame of reference and the frame of reference the imagined observer would have, in favor of the imagined frame of reference.[2]

2. Simons and Wang (1998) have argued that adjusting for viewpoint changes when observers actually move around an array depends more on their use of information from visual flow, proprioception, and efference (i.e., on dead reckoning) than on their use of the external framework of landmarks. We would argue that both sources of information are probably important in this situation. However, whatever the case when people actually move, when movement is only imagined, as in the perspective-taking task, dead reckoning clearly cannot be used in answering the question.

The conflict between actual and imagined frames of reference created by the classic perspective-taking task is not a necessary part of testing the ability to work out spatial relations from another point of view. The hypothesis that conflict between frames of reference accounts for young children's difficulties with the classic perspective-taking task has received empirical support from demonstrations that different ways of asking the perspective-taking question lead to much better performance. Huttenlocher and Presson (1979) used what they called "item questions" instead of a selection task. That is, they asked questions about what object would occupy a specified position with respect to another object, instead of using picture or model procedures. For instance, one can say, "If you were over there, which toy would be closest to you?" Huttenlocher and Presson (1979) found that 9-year-old children had little difficulty answering item questions about other perspectives, even though they showed the usual problems with the classic task.

Nine-year-olds are, however, close to the age at which children generally become able to solve the classic task as well. Hence it might be argued that item questions are merely a slightly easier way of asking perspective questions, but that perspective-taking ability still rests on late-appearing forms of spatial representation. In contradiction to this argument, Newcombe and Huttenlocher (1992) demonstrated that children as young as 3 years showed performance on item questions that was substantially above chance levels (70 percent correct, compared to 25 percent for chance). The fact that 3-year-olds did so well on a task not solved by 9-year-olds in the classic format suggests strongly that Piaget was wrong about late developmental change in spatial representation. Preschool children seem to have relatively little difficulty with representing the location of objects and with retrieving locations given specific reference points. That is, they not only know that others have a different view than they do, they can work out the spatial relations seen from the others' vantage point under suitable questioning.

Let us not reject the idea of qualitative representational change without considering potential counterarguments. Piagetian theorists (e.g., Chapman 1988) generally deny that demonstration of younger-than-expected ability to perform a certain task is a fundamental challenge to Piaget's claims. Chapman (1988) defends Piaget's analysis of perspective taking by focusing in particular on Borke's (1975) experiment showing that young children can solve the Three Mountains task if they are asked to indicate the perspective of another observer simply by rotating a movable display. Chapman argues that this fact is unsurprising within Piagetian theory, since the physical structure of the movable display holds constant the internal relations among the objects in the display. The original task was explicitly designed by Piaget to explore if children could coordinate and transform such information, and the turntable obviates the need for this processing. Chapman's criticism of what Borke's study shows about spatial processing seems to us correct (although her work does seem to undercut Piaget's claim of egocentrism, which is a different matter). What would a theorist of Chapman's persuasion say, however, about the evidence of early success with item questions? We have argued that these studies bear on issues of spatial representation and question Piaget's characterization of developmental change. We imagine such a theorist arguing one of three things to explain our results within the Piagetian framework.

First, it might be said that very early spatial coding is still topological, although developing into projective representations by 3 years rather than 9 or 10 years. Thus, such a critic might argue, only the trivial issue of age norms is being addressed, not the more important description of developmental sequences. There is a major problem with this defense, though, in that it does not explain why the item questions should produce such different levels of performance than the classic formats. If children older than 3 years have projective representations, they should not find choosing among alternative pictures difficult. Our approach does explain the difference between performance with the two types of task.

Second, a Piagetian theorist might propose that item questions can actually be answered using topological coding. This proposal does not deal, however, with logical reasons to doubt the usefulness of the topological, projective and Euclidean typology (e.g., Mandler 1983) or with a key problem with Piaget's definition of topological coding (Newcombe 1988). "Topological" representation is generally said to include the concept of neighborhood or proximity. In fact, without the inclusion of neighborhood and proximity, describing children's spatial understanding as "topological" is clearly false, since children younger than 9 years are in fact able to remember the location of objects even when those objects are near, but not coincident with, referents, as when a bunch of keys is dropped a foot or so from a chair (e.g., Acredolo, Pick, and Olsen 1976). Yet incorporating the notions of neighborhood and proximity as part of topology makes the use of the term topology formally wrong (in terms of a mathematical definition). More important, such incorporation causes internal contradiction. It is hard to see how neighborhood or proximity could be defined in the absence of a notion of metric extent, something Piaget explicitly denied to the young child (Newcombe 1988). Postulating topological relations not only between physically present entities but also between an imagined observer and an object hidden under a cover, as is required for a topological analysis of answering item questions, seems to strain the notion of topology even further.

Third, it might be suggested that answering item questions is part of practical rather than conceptual space, a key Piagetian distinction. However, we find it difficult to see item questions as practical. Item questions do not involve action in the world, or reactions to direct perceptual feedback about closeness to goals, looming objects requiring detours, and so on. Item questions involve the combination of perceptually available information and information stored in memory to yield new information. Thus we would argue that the simplest account of development in perspective taking, and the one most consistent with other evidence on early spatial coding, is that some kinds of perspective-taking questions are difficult because they require dealing with conflicts between frames of reference.[3]

Although task simplification leads, in many cases, to assessment of different cognitive skills than were tapped in Piaget's original tasks, the item-question findings are, we think, a different matter. They address the issue of spatial representation and require substantial revision of thinking about perspective taking. The fact that children can correctly indicate another's viewpoint in certain circumstances supports a model of spatial coding and transformation which is quite different from the topological-to-projective/Euclidean hypothesis. By the age of 3 years, children, like adults, seem to code the location of objects with respect to an external frame of reference which interferes with answering certain kinds of questions about the array—those that require using a conflicting frame of reference. This approach has proved its worth by generating testable hypotheses which have turned out to be empirically supportable (Huttenlocher and Presson 1973, 1979; Newcombe and Huttenlocher 1992; Presson 1980, 1982).

Given this analysis of perspective taking, one may ask how children eventually do become able to solve the classic perspective-taking task. A likely possibility is that they learn to use a strategy on this task that helps to overcome the conflict. Specifically, if one focuses on a single item in an array of items, and asks oneself an item question about it (i.e., "If I were over there, what item would be in front?"), one can then use the answer to this self-posed question to pick the correct picture or

3. Lynn Liben has suggested several times in conversation that conflict between frames of reference is exactly what Piaget was interested in when he devised the perspective-taking task. While close reading of Piaget suggests that he was aware of this component of the task, he certainly also suggested several other ways of conceptualizing it. Perhaps the best way to think about all the papers inspired by the Three Mountains task is that they allowed for the unravelling and clarification of the many possible interpretations of children's performance considered by Piaget.

model. Bialystok (1989) found that 7-year-olds used such a strategy only to a very limited extent but that it was consistently apparent by ages 9 or 10.

The present analysis does not deny that an important transition takes place in late childhood in perspective taking. The ability to deal with conflicting frames of reference using strategies is important in real life, and it is apparently not present until toward the end of elementary school. The ability is central to performing important spatial-cognitive tasks, as when an architect imagines how a building will look from a variety of perspectives and draws pictures of these views. Adults, unlike young children, are able to cope with such situations. Although even adults take longer to answer classic perspective-taking questions than to answer item questions (Presson 1982),[4] dealing with conflicting frames of reference becomes possible, even if challenging. From a functional point of view, adults and young children are quite different. Architects are happy to be able to use computer-assisted designs to ensure swift and accurate visualization, but they did design buildings before such aids were available.

Summary
What develops with age on the perspective-taking task is not a fundamentally different form of spatial representation. Instead, what develops is a strategy or set of strategies for dealing with a cognitive situation challenging to individuals of all ages: the situation in which two or more frames of reference must be maintained at once, yet distinguished from each other and interrelated. The advent of success on such tasks in late childhood is a functionally significant transformation of cognitive ability.

Children's Judgments of Distance

The Findings
A child of five years is shown a small display containing two objects, perhaps a toy house and car. Asked whether the distance between the house and the car is small or large, she says "small." (Actually it doesn't matter whether the child says "small" or "large;" the point is for her to produce some assessment of the distance.) Then a model tree is placed between the house and the car. Asked whether the distance between the house and the car is smaller than before, larger than before,

---

4. This difference is not the result of answering questions simply being easier in general. Adults take longer to answer questions about an item's position (position questions) than to answer questions about what item is in a specified position (item questions).

or the same as before, the child says "smaller than before." By contrast, an older child or an adult would unhesitatingly judge that the distance remained the same.

A similar developmental change is seen in how preschool children and older children answer questions about the length of an object. A child is shown two sticks of the same length, with their ends aligned. Even young preschoolers readily agree that the sticks are the same length. Then, one of the sticks is moved over on the table, so that its ends no longer align with the ends of the other stick. Younger, but not older, children generally say that one of the sticks is now longer than the other, focusing on the ends of the display and pointing out that one of the sticks protrudes farther than the other.

Preschool children have difficulties with other questions involving distance too. For instance, when children are shown race cars that start and stop a race at the same or different points and go the same or different speeds, they often have trouble evaluating distance, speed, and time separately. Young children usually focus simply on the end points in judging distance or speed.

Piaget argued that children's difficulties with questions about distance and length indicate a momentous lack, a deficiency entailing being fundamentally unable to encode spatial relations. "Where this concept is present, children will view space as a common medium containing objects and spatial relations between objects. . . . When it is absent, space as a medium common to several objects will be uncoordinated" (Piaget, Inhelder, and Szeminska 1960, p. 69).

The Controversy

As we have already seen in the past two chapters, there is reason to question Piaget's claim that young children are unable to encode the distances between objects and salient environmental landmarks and frameworks. By 16 months, children show a robust ability to encode the location of an object buried in a sandbox, digging in reliably differentiated locations based on their observation of where an object was hidden (Huttenlocher, Newcombe, and Sandberg 1994). By two years, children use distant landmarks to help track their position in space and to locate objects (DeLoache 1984; Mangan et al. 1994; Newcombe et al. 1998). These demonstrations indicate that children much younger than five years or so (the age when children are still making mistakes on the conservation of distance and length tasks) are indeed able to encode distance and spatial relations.

Children's ability to encode distance in search tasks may, however, not be relevant to their ability to represent distance information and use it to solve problems. Piaget described visual estimation as percep-

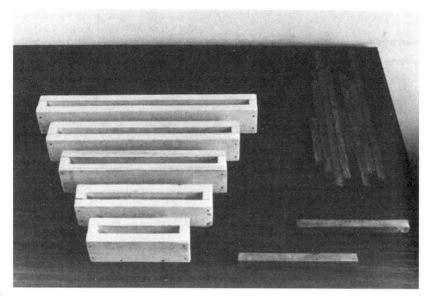

Figure 5.6
Materials used in the Schiff (1983) study. A pair of equal-length sticks is shown in the transformed position. Sticks were actually located farther from the boxes while the study was being performed.

tual or "figurative" knowledge, claiming that such knowledge is different in kind from the conceptual knowledge required for judgments and reasoning. Thus Piaget would argue that when children succeed in finding objects in the sandbox, or are helped in locating objects by the availability of landmarks, they are demonstrating a practical or a perceptual knowledge rather than a conceptual understanding. Although there may be truth in this argument as it concerns the first years of life, as indicated in the first section of this chapter, recent investigations have undermined the argument that preschool children's answers to questions about distance and length indicate a fundamental lack of appreciation of space as a structuring medium (Bartsch and Wellman 1988; Fabricius and Wellman 1993; Miller and Baillargeon 1990; Schiff 1983).

In Schiff's (1983) investigation of conservation-of-length tasks, preschool children were asked to look at a stick and to indicate which of five boxes the stick would fit into, as well as answer standard verbal conservation questions. Children performed these tasks both before and after the sliding transformation which usually causes nonconservation judgments in children of this age (see figure 5.6). Almost all children gave nonconserving responses on the standard verbal task.

But almost all of these "nonconserving" children correctly judged that a stick would go into the same box after it was moved as before it had moved. This indicates an understanding of the basic principle of conservation: that the quantity of length is unchanged by a transformation.

The children's judgments did not seem to be perceptually based, for two reasons. First, the children often picked a box not the right size for the stick, showing relatively poor perceptual estimation; the key fact was simply that children did not change their judgments when sticks were misaligned. Sticking to the same box shows that on a conceptual level, the children did not think sliding the stick across a table changed the physical dimensions of the stick. Second, children's judgments of which box would fit which stick stayed consistent after sliding transformations, even when the array of boxes was composed of choices with lengths so similar that they were not easily perceptually discriminable.

Miller and Baillargeon (1990) worked with conservation of distance rather than length. As had Schiff, they introduced a concrete referent for "same." They showed children two blocks a certain distance apart, and asked the children to pick a stick that would bridge the distance between the blocks. They then interposed a screen (of various widths, heights, and orientations) between the blocks, and asked the children to pick a bridging stick again (see figure 5.7). Children as young as 3 years did significantly better on the bridge task than on the classic conservation task (see figure 5.8).

As with Schiff's box task, the advantage of the bridge task over verbal responses did not seem to be based on visual estimation. The children were actually quite bad at estimation, and at no age did they do better than chance at picking the correct stick. The important finding was simply that the children did not change their judgment of which stick would bridge the gap as a function of an object being interposed between the two blocks. This phenomenon was conceptual rather than perceptual. Children stuck to their choice of bridge, despite the fact that other children of the same age, shown the same displays of two blocks with and without interposed objects, estimated the distances between blocks as shorter when interposed objects were present than when they were not. Thus the maintenance of the "same bridge" response occurred in the face of perceptual appearance of a shorter distance with an interposed object. Children's correct judgments did not depend on perception; they occurred despite perception.

The findings of Schiff and of Miller and Baillargeon indicate that preschool children's difficulties with conservation of distance and length tasks may be due, at least in substantial part, to nonspatial factors. But what are these nonspatial factors? Schiff suggested that

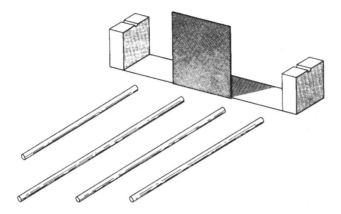

Figure 5.7
Apparatus used in Miller and Baillargeon (1990).

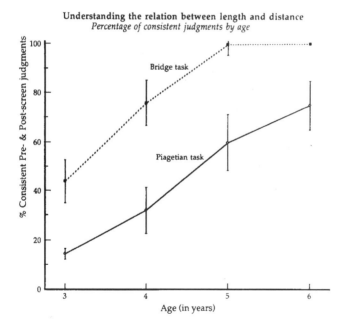

Figure 5.8
Percentage of consistent judgments on bridge task and on standard Piagetian task. From Miller and Baillargeon (1990).

children think that the words "longer" and "shorter" refer to end point comparison, that is, that their problem is one of mapping words onto the correct concept. Similarly Miller and Baillargeon suggested that children have difficulty sorting out the various meanings of the terms "near" and "far" used in their study. Although Miller and Baillargeon's experiment 2 showed that preschool children have some understanding of the terms, in that the children correctly judge that the blocks were nearer when the experimenter moved them closer together and farther apart when the experimenter moved them apart, Miller and Baillargeon argue that there are many other senses of "near" and "far" which are more difficult to understand. By comparison with verbal questions using extent terms, the "which box?" or "which bridge?" tasks offer an unambiguous criterion for what the experimenter means by "same" or "different."

From a Piagetian point of view, however, these findings do not allow the conclusion that 3- and 4-year-olds have concepts of distance and length (cf. Chapman 1988; Fischer and Bidell 1991). Such theorists would argue that children do better on the box and bridge questions than they do on questions using extent terms because of the concreteness of childhood thought, highlighted by Piaget, and the dependence of children's thought on action in specific well-defined situations. The box and bridge tasks pose concrete, action-linked questions. They refer to very specific states of affairs with which the children may have considerable personal experience. When building block structures or putting away toys, children probably have plenty of occasion to observe the situations under which objects do and don't fit across gaps or into spaces, and to see when these fits do and do not change. In this view, the relative inability of young children to induce a stable abstract spatial meaning for terms such as "longer" and "shorter" or "near" and "far," is a key limitation on early thought, not an epiphenomenon of the complexity of our systems of linguistic reference. The perceptual-conceptual distinction may not be quite right, such defenders would argue, but the "concept" of distance shown by 4-year-olds in these studies is still not the concept in adult terms: it is tied to concrete and action-based situations.

A third demonstration of early competence on distance-judgment tasks undermines this defence, however. Bartsch and Wellman (1988) studied preschool children's understanding of distance in situations in which children could not rely on perceptual estimation, because displays were screened, and in which there was no concrete object, such as a box or a bridge, used in questioning to support children's judgments. Bartsch and Wellman studied two principles of distance judgment. One, which they termed the "direct-indirect principle," is the

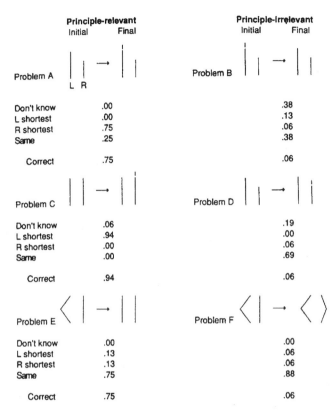

Figure 5.9
Problems and data from Exp. 2 of Bartsch and Wellman (1988).

idea that the shortest distance between two points is always the straight one. A second, which they termed the "same-plus" principle, is the idea that if two routes are the same length initially, but one is then added to, the one added to must be longer. Preschool children in Bartsch and Wellman's study saw initially equal lines, with one line then transformed behind a screen, either by being bent and stretched or by having a portion added or subtracted (see figure 5.9). The children correctly judged which line was the shorter. They were not reacting simply to the transformations themselves, using a rule such as "If you add to one line, that line is longer," because when the lines were not initially equal and thus the correct answer after a transformation was indeterminate, the children were not able to determine which was longer (as, indeed, neither would an adult be able). Thus their decisions were governed by logical principles. In a subsequent study Fabricius and Wellman (1993) added to these findings by showing that 4-year-old children also used the direct-indirect principle to judge relative distances with a much larger-scale display. While this display was unscreened, there was evidence that judgments did not depend on perceptual variables such as the time needed to scan the two paths.

Another line of research on children's understanding of distance concerns the relations of such understanding to children's concepts of time and speed. At 10 years, children can integrate speed, time, and distance in a metric fashion (Wilkening 1981, 1982). However, while children of 4 to 6 years appreciate that more time and more speed both imply more distance, they are confused about the relation of time and speed (C. Acredolo, Adams, and Schmid 1984; Matsuda 1994). Distance often dominates the other two dimensions. For instance, Siegler and Richards (1979) found that 5-year-olds are likely to use the "end-point rule" identified by Piaget (i.e., the distance is the same if the objects end up in the same position), with transition to a true distance rule occurring between 5 and 11 years. However, Richards (1982) reported that many 5-year-olds initially using an end-point rule to judge distance could be brought to use a true distance rule through training.

Overall, several literatures converge to suggest that by the age of 3 or 4 years, children show substantial ability to judge problems involving distances. The conclusion that these findings demonstrate early competence should not, however, be overdone. Children's abilities to perform these tasks are clearly not at adult levels. For instance, figure 5.8 shows considerable developmental increases in accuracy on Miller and Baillargeon's bridge task. One interpretation of these increases is that Piaget was right about the existence of developmental changes in understanding of distance but simply wrong about the ages at which steps in the sequence are reached. It may be argued that revision in ages is relatively inconsequential theoretically; presumably this is what Piaget meant when he said that age is the "American question."

We believe what is at stake here goes beyond the issue of age norms, however. What is most important about these studies of early distance and length concepts is not the mere demonstration of earlier competence than Piaget imagined. Such a demonstration is clearly subject to the criticism that what has been demonstrated is nonchance performance rather than adult ability. Rather, what is important is that the studies suggest an alternative picture of the nature of developmental change. Specifically, change does not seem to be from practical to abstract understanding, since early competence can be seen equally in situations in which practical skills may be drawn upon, as in the box and the bridge tasks, and situations in which understanding seems more abstract, as in Bartsch and Wellman's experiments. Nor does change seem to occur from topological to projective or Euclidean representational formats. Instead, it seems likely that developmental change occurs in the mapping of conceptual onto linguistic terms (an issue to be considered in more detail in chapter 7), in the ability to reconcile and coordinate perceptual and conceptual judgments when these are in apparent conflict (e.g., as when a gap between two blocks looks smaller than it did before an object was interposed, but one's understanding of space leads to the judgment that the same bridge is needed to cross the gap), and in the ability to use measurement operations to assess and compare distance objectively. Miller (1984, 1989) has emphasized the Vygotskyan point that measurement tools and routines can serve as a "tool for thought" which propels as well as reflects cognitive development. Ellis and Siegler (1995) have shown that acquisition of measurement procedures in the early elementary school years is heavily influenced by children's competence at this age in using counting to assess quantity. For instance, this reliance on counting led children to be resistant to the idea that they could measure length with any instrument not divided into units. In sum, by the age of 3 or 4 years, children have simply begun what will end up being the long road to being able to reason about distance precisely and consistently.

Summary
There is little reason to think that 3-year-old children lack a basic appreciation of space as a "structuring medium." Changes in conceptual-linguistic coordination, in the ability to deal with conceptual-perceptual conflict, and in the correct use of measurement tools and operations likely account for age-related improvement in children's ability to make judgments about distance. These changes are themselves important, however, and have substantial consequences for spatial thought. For instance, one can discuss the consequences of taking certain routes much more straightforwardly with an adult than with a preschooler.

Children's Way-Finding and Route Representations

We turn now from aspects of spatial thought upon which Piaget based his arguments for developmental transformation to consider tasks of large-scale spatial cognition and way-finding. A variety of means can be used to code location and hence to plan routes. Movement can be planned and monitored either with reference to landmarks as continuously available "beacons" or by using landmarks on a more periodic basis to assess and correct a route encoded from kinaesthetic and proprioceptive feedback about length and direction of movement.

When a desired goal is distant, reaching it may involve planning a route in which there are subgoals. That is, for example, getting to a blueberry meadow may involve reaching a climbable approach to a mountain, which may involve following a brook upstream, which in turn may require finding the brook from one's current location. Route planning in the large-scale environment occurs in situations in which it is difficult to derive an overall perspective, that is, to obtain a bird's-eye view. When one's view of the environment is physically restricted, relevant landmarks are seen sequentially rather than simultaneously, and relating them requires inference from information about the routes which link them, from the known interrelation of intermediate landmarks, from the known relation of the regions in which they are located, and so on. In fact the need for such inferential linkage may be considered definitional of large-scale spatial cognition.

Knowledge of the sequence of landmarks which must be followed to reach a goal has been termed *route knowledge*. One approach to route knowledge has been to claim that it constitutes a special kind of spatial knowledge, intermediate between simple knowledge of landmarks and survey knowledge, or an integrated overall representation of the interrelation of landmarks (Siegel and White 1975). Siegel and White suggested that people begin learning about large-scale space by learning landmarks in a new area, and then begin to encode the order of landmarks which demarcate routes from specific starting places to salient goals. At some point, when enough routes and landmarks are encoded and interrelated, overall configurations of space (or survey knowledge) are formed.

Siegel and White portrayed the sequence of landmark to route to configurational knowledge as typical of adults learning a new environment, and also as characterizing the sequence observed in development. They argued that younger children showed lower levels of spatial knowledge than older children and adults, namely, that while young children could notice landmarks and form route maps, they did not form integrated spatial representations. This difference was said to be linked to children's lack of an objective system of reference: "landmarks and routes are formed into clusters, but until an objective system of reference is developed, these clusters remain uncoordinated with each other" (Siegel and White 1975, p. 46).

We argue, however, that, after the first few years, developmental changes in spatial representation consist in refinement of a basic system of spatial location, rather than a transformation. Thus, for instance, while children do not begin to use category prototypes in correcting estimates of location for both dimensions in a two-dimensional encoding situation until after the age of 7 years or so (Sandberg, Huttenlocher, and Newcombe 1996), such change simply further reduces variability in estimations that were already centering on the correct location. From this perspective, children's knowledge of the large-scale environment

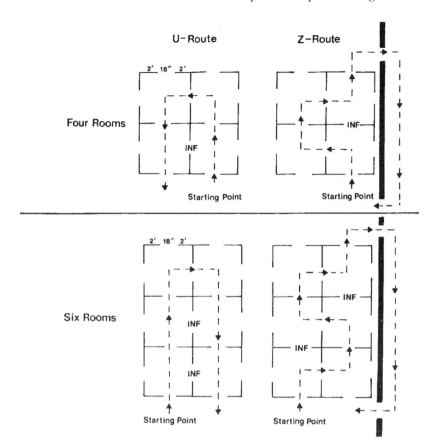

Figure 5.10
Maps of the two routes in the four- and the six-room layouts analyzed by Hazen, Lockman, and Pick (1978). Solid lines stand for walls, spaces stand for doors, and the dotted lines show the route. INF stands for the doors at which children were asked for inference judgments.

is likely to be route based only in circumstances in which the inference necessary to achieve integrated spatial representations is overly taxing.

Let us examine some of the evidence for the position that young children's representations are route based and fragmented. One well-known study on this topic was conducted by Hazen, Lockman, and Pick (1978). Hazen et al. worked with 3-, 4-, 5-, and 6-year-olds who were taken on a path through a series of small inter-connected rooms, each with a distinctive animal in the middle (see figure 5.10 for the two configurations and two routes they used). Children followed a path repeatedly, always in the same direction, until they were able to anticipate which animal followed which without error. They were then given several tests of spatial knowledge. On one, route reversal, they were asked to traverse the "house" along the same path but in the opposite direction. Even 3-year-olds did this task excellently, a fact that is congruent with evidence already reviewed that young children can encode the direction and extent of their own movement and use that information to engage in shortest-route behavior. Children were also asked to anticipate what animals would be seen along the reversed path (the landmark reversal task). Again, even 3-year-olds did excellently, showing that they had a memory for the route sequence that could be operated upon to read off in the opposite order from that experienced. On a third test, however, children were asked inference questions, shown in figure 5.10, involving what would be seen if they went through doors never directly experienced. They did quite poorly on these questions, and performance improved with age. Poor performance was also seen on a fourth test, of children's ability to construct models of the house.

Hazen et al. concluded that young children's spatial representations are "route-like and poorly integrated." Pick (1993) reiterated this basic conclusion, adding that the model-building assessment in the Hazen et al. study may have overestimated the extent of children's spatial knowledge, because the children may have constructed the knowledge during the course of building the model, making inferences "on line," rather than simply externalizing an existing representation.

We would interpret these data somewhat differently. First, as Hazen et al. recognize, poor performance on model construction is likely to involve many factors of planning and understanding of model conventions; it is not a direct way to assess spatial representation (cf. Kosslyn, Heldmeyer, and Locklear 1977). For instance, one child in the Hazen et al. study who modeled the rooms as arrayed in a single line nevertheless indicated when turns would be taken with a finger motion as she showed the route, an idiosyncratic convention which nevertheless indicated her underlying knowledge. Second, and more important, poor performance on the inference task may reflect the fact that in this study there were no visible external landmarks when children were in the house. Thus, encoding could only be with respect to the order of the animals distinguishing the rooms, and the direction and distance of the children's own movement. Answering inference questions would require integration of these sources of information, so the order of the animals would be laid out on a mental model of the route, a model that would then allow for "reading off" the answer to the inference question. However, such a task of mental model construction is very challenging, even more difficult than the task of physical model construction, since each component has to be maintained mentally.

In an environment in which objects (e.g., the animals in the Hazen et al. study) could be coded with respect to the external framework (e.g., would occur if the "house" in that study had no "roof" and hence children could see the wider environment of the lab), children's ability to answer inference questions might have been much better. The relation of the animals to the environmental landmarks would allow for much easier construction of a mental model. In fact, arguably, such questions would no longer tap inference at all, since the answers would be directly represented.

The point we are trying to make here concerns what should count as evidence about the nature of spatial representation. It may well be the case that younger children find it more difficult to construct integrated representations of particular large-scale spaces, but we do not think this is because their idea of space is fundamentally different from that of older children or that their coding differs vastly. Rather, individuals of varying ages differ in their ability to reason about the kind of fragmentary information that is frequently all that is physically available in certain environments.

Age differences in ability to integrate spatial information may have several causes. One possible cause is differences in processing capacity, a factor stressed by Siegel and White (1975) along with their idea that young children did not have an objective frame of reference. Processing capacity is notoriously difficult to define in a noncircular way (see Flavell's 1984 comments on Case's approach to development), but it has some appeal nevertheless. There is evidence, as reported in chapter 4, that children's difficulty in forming and using categorical representations on two spatial dimensions simultaneously through the age of 9 years depends on some kind of capacity limitation (Sandberg 1995; Sandberg, Huttenlocher, and Newcombe 1996).

A second possible cause of difficulty in inference is that, as we have seen, younger children code location more variably than older children, in that they use larger categories for coding on the categorical level, and fail to use two categories in two-dimensional coding. Greater variability leads to greater potential for large errors in estimation, and large errors in estimation concatenate in the inference process so that inference may become impossible. Findings on the development of children's ability to point to landmarks not visible from their current position indicate that progressive improvement in this ability occurs through the age of 12 years or so (Anooshian and Nelson 1987; Anooshian and Young 1981; Cousins, Siegel, and Maxwell 1983).

The ability to draw inferences and to construct mental models when locations have not been encoded with respect to a common frame of reference is thus an important line of developmental improvement. Other kinds of developmental change, falling into the category of memory strategies, also affect children's ability to find their way around in the world. Simple memory strategies have been demonstrated in toddlers (DeLoache 1984) but are well known to increase with age. Changes in strategic aspects of the way-finding process include age-related improvement in selecting salient landmarks and in noticing landmarks at turns (Allen et al. 1979) and acquisition of ability to use a strategy in which one looks around at turns to encode the perspective that will be seen on the return trip (Cornell, Heth, and Rowat 1992). Changes of these kinds lead to definite improvement in such practically important tasks as the ability to reverse a route taken once around an unfamiliar area, or to point to a direction to go when off route. Improvement in these situations have been seen to occur between the ages of 6 and 12 years (Cornell, Heth, and Alberts 1994; Cornell, Heth, and Broda 1989; Cornell et al. 1992).

Summary
There is little evidence that young children lack an objective frame of reference or are inherently unable to construct integrated spatial representations. They may, however, find it more difficult than older children to integrate their spatial representations when a common frame of reference is not available, and fragmented spatial relations have to be related by inference and by the construction of mental models. They are also less likely than older children to use effective strategies of landmark selection and route examination to help them in navigating in unfamiliar areas.

Children's Logical Searching and Planning

Investigators have used spatial search behavior to examine a number of questions about development in addition to the nature of spatial representation, notably the development of logical processes and of planning ability. We examine each of these two topics briefly here, summarizing the key findings and discussing how the results enhance our understanding of spatial behavior.

Logical Search
Studies of logical search involve an experimenter showing a subject that something exists at a certain point along a route, and then showing the subject that the object is missing at a later point. For instance, Wellman, Somerville, and Haake (1979) had 3-, 4-, and 5-year-old children participate in a series of games, one at each of eight locations marked on a familiar preschool playground. At one location a camera was used, and at a subsequent location it was found to be missing. The

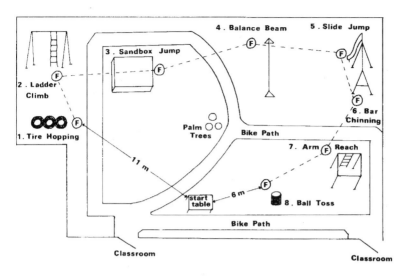

Figure 5.11
The playground used in the Wellman, Somerville, and Haake (1979) study. The children performed eight tasks at eight locations.

question was whether, when children were asked to search for the camera, they would restrict their searches to the area between the location at which it was last seen and the location at which it was first seen to be missing.

From the point of view of spatial representation, what is required to succeed at this task is the ability to remember the sequence of locations visited. Although representation was not directly assessed, it seems very likely that children as young as 3 years would be able to encode such a sequential route representation (Hazen, Lockman, and Pick 1978). This supposition is strengthened by the fact that encoding of the sequence would be facilitated by the layout of the playground, since locations followed each other in a simple order (see figure 5.11).

If children had encoded the route, the question of the study concerns the development of the second ability required to succeed at this task, namely knowing that the space between where the object was last seen and where it was first missing should demarcate the search space. In fact, children as young as 3 years seem to appreciate this fact as well as older children, a finding obtained in a variety of spaces and with a variety of procedures and controls (Anooshian, Hartman, and Scharf 1982; Haake, Somerville, and Wellman 1980; Wellman et al. 1979). In a somewhat simpler situation, children as young as 2 years have also been found to show logical search ability (Haake and Somerville 1985). Haake and Somerville traced this logical ability to its roots, finding that infants demarcated search areas at above-chance levels by 15 months, and quite consistently by 18 months.

The milestone documented by Haake and Somerville coincides neatly with Piaget's observations of infants. The difference between the infant in Piaget's stage 5 of object permanence (i.e., an infant who does not make the stage 4 A-not B error) and the infant in stage 6 (attained at around 18 months) is that the latter, but not the former, is able to pass a task in which an object undergoes invisible displacement (i.e., in which the object is last seen enclosed in a hand disappearing under a cloth, and the closed hand then traces a path to another cloth, finally opening to reveal the object missing). Piaget thought this change showed something about object permanence and coding of spatial location. However, given evidence that infants understand the permanence of objects in the first year, and that they can also code location with some accuracy in the first year, the ability to succeed at tasks of invisible displacement appears to show something else about development. Specifically, in the second year of life, infants seem to develop a new ability to reason logically about spatial-temporal sequences.

Planning Shortest Routes
From the point of view of biological adaptation, a central purpose of representing the environment is to allow for the planning of routes to get to desired objectives. But when someone has more than one objective in mind at a time, it is desirable to know not only the location of the two objectives but also how to plan a route to reach each of the two objectives that minimizes the total distance to be traveled. Minimizing distance may not be too important if there are only two objectives and when these objectives are restricted to a small general area. However, as numbers of goals increase and the distances between the goals also increases, the conservation of energy expenditure achieved by planning shortest routes becomes an important adaptational goal. Studying shortest-route behavior is the central concern of many students of foraging behavior.

As with logical search, being able to perform a spatial planning task requires abilities in addition to the ability to represent spatial location, specifically, the mental construction and comparison of alternative possible routes. The demands of such construction and comparison clearly increase with the number of steps (in this case, locations) involved in the planning process (Klahr and Robinson 1981).

Research on spatial planning was reviewed by Wellman, Fabricius, and Sophian (1985). The evidence suggests that children as young as 1 or 2 years may engage in some simple forms of planning shortest routes. For instance, Lockman and Pick (1984) found that by 18 months,

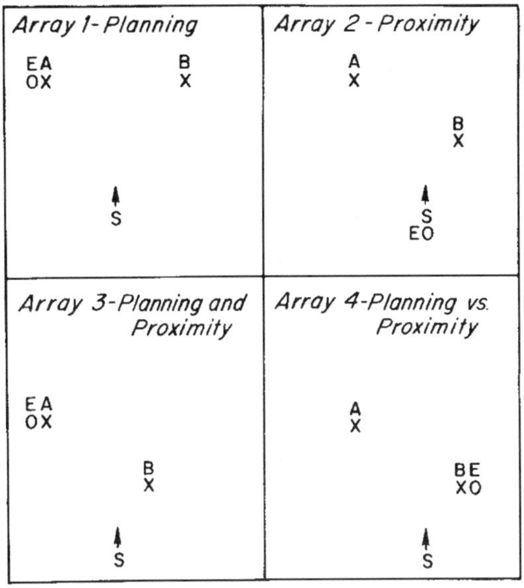

Array 1- Planning	Array 2 - Proximity

EA   B
OX   X
	A
	X
	
	               B
	               X
	
	        ↑
	        S
	        EO

Array 3-Planning and Proximity	Array 4-Planning vs. Proximity

EA
OX
 	
          B
          X
 	
     ↑
     S
	A
	X
	
	               BE
	               XO
	
	     ↑
	     S

Figure 5.12
Arrays used to study children's planning of spatial routes, from Wellman, Fabricius, and Sophian (1985). S is the start location, A and B are the toys children are to gather, and E is the end point.

infants positioned toward one end of a long wall separating them from their mothers would go the shorter rather than the longer way around the wall to reach their mothers. Wellman et al. argue that this behavior, and most spatial planning through the age of 3 years, depends on the *sighting* of particular desired goals and the obviously greater proximity of one means of reaching that goal. Given that the mother is visible, in the Lockman and Pick situation, and that the two possible routes can also be directly perceived, with one end of the wall closer than the other end, the child is not required to build and compare mental models to succeed at the task. However, other findings indicate an early ability to compute shortest routes, when the goal is not visible (e.g., Pick and Rosengren 1991; Rieser and Heiman 1982).

Planning to take a shortest route and to minimize distance traveled seems to occur in conflict situations in which one must first visit a location not obvious in direct sight, or not proximal to the starting point, while avoiding initially visiting an object in sight or proximal. Ability to plan routes in conflict situations is not robustly evident until 5 or 6 years. Wellman et al. have conducted research in which planning requires not going first to the closest of two goals (see figure 5.12). In this situation, given the location of the end point, going first to the closest goal will result in a longer path than going first to the distant goal, then to the goal closer to both the starting and the ending points. Children of 5.5 years could succeed at taking the shortest route in this conflict situation, but children of 4.5 years could not solve the problem. The difficulty the younger children have in this situation may stem from problems with logical analysis of problems with inhibiting a dominant or prepotent response.

Summary

The earliest origins of the ability to delineate a plausible search area for a missing object appear to lie at about 18 months. This age is also the first at which children have been reported to be able to compare simple routes. Planning of shortest routes in conflict situations requiring the inhibition of dominant response tendencies, by contrast, is not evident until toward the end of the preschool years.

What Kind of Theory of Spatial Development Is This?

In this chapter we have examined the possibility that spatial representation, as opposed to spatial coding, evolves in the second year of life, and we have criticized nativist claims regarding spatial representation. We then analyzed the development of children's ability to perform four tasks that require reasoning about spatial representations. In each of these four cases, we have found evidence of essential competence dating back to 3 or 4 years and of subsequent developmental transition along several lines. Children invent strategies for dealing with situations in which they must construct models with regard to one frame of reference while being situated within another frame of reference. They acquire mappings of linguistic onto conceptual categories, which aids in discussing space. Finer subdivision of space into mentally imposed categories improves the accuracy of spatial coding, easing the process of inference. Inference may also be aided by increased capacity to construct mental models. Way-finding strategies are picked up, including the ability to pick distinctive landmarks, and tricks such as turning around on occasion to visualize a route from the perspective to be seen on the return trip. Children are able to plan on minimizing travel in situations in which conflicts exist between various impulses.

What kind of approach to spatial development is being advocated in this description? We have argued against nativism. However, a nativist counterreaction might be that everything important in our analysis is present from an early age; even the advent of more durable memory for spatial information might be dismissed as a neural hardware problem. What develops, in this view, is relatively trivial (Spelke and Newport 1998). This view of development rests crucially, however, on how one defines what is important. Clearly, development must depart from some foundations; no one has advocated a *tabula rasa* for quite some time. If one started development from different foundations, one might well get a different product or no product at all, and in this sense a nativist position has a certain amount of merit. But one might also get a very different product, or no product at all, with different environmental input or without the addition of durable coding or logical reasoning abilities. In this sense radical environmentalism might be said to also have a certain amount of merit. In sum, we do not believe that the picture of development presented in this chapter can be said to be an essentially nativist one.

The approach also differs in many respects from the traditional Piagetian portrait of spatial development. It potentially agrees with that account, however, in one very interesting way. Between the age of 1 and 2 years, we believe there are at least some reasons to suppose that children acquire more durable memories for spatial information, capable of supporting delayed action plans and of being used in inferential reasoning. Such a transition is, broadly speaking, what Piaget had in mind when speaking of the advent of symbolic thought. Subsequent to this period, however, we part company with Piaget. Children of three years do not seem to have qualitatively different methods of spatial coding from older children. There is evidence that they possess the ability to represent space with respect to both external and self-based frames of reference. What develops, then, are abilities and strategies for dealing with the complexities of real-world space.

A traditional Piagetian reaction to this analysis might be that it commits the usual crimes of simplifying Piaget's tasks to find better performance at younger ages, failing to challenge Piaget's views because it has not really addressed them at all. We have tried to address this criticism in detail throughout the chapter. Another reaction from a traditional Piagetian might be that with the possible exception of our speculation of a symbolic shift, we endorse an additive rather than a constructivist view of development, in which development is quantitative rather than qualitative. By the age of 2 or 3 years, they would say, you see children as having acquired the basic building blocks of spatial representation, after which they just add a few strategies or a bigger dollop of processing capacity, rather than undergoing fundamental transformation.

This reaction ignores, however, the demonstration from dynamic-systems theory and from some connectionist models that quantitative changes of various kinds can be associated with qualitative changes in system functioning (Elman et al. 1996; Thelen and Smith 1994). In order to be able to actually build anything of interest from a foundation, children need also to acquire quantitative increments of mortar and brick, and to build up the skills of the brick-laying trade. Without these, they cannot create a house, surely a structure qualitatively different from a foundation. In the spatial realm it seems that children are acquiring strategies, capacity, linguistic terms, and so on. All are admittedly quantitatively incremental, but together they pass over important functional thresholds, enabling qualitatively different behavior than that usually seen in younger children.

The description of development offered in this chapter is thus broadly compatible with the constructivist approach to cognitive development. Children actively work to develop effective strategies for spatial problem-solving, building on a strong basis of spatial encoding ability developed in the first two years of life. Russell (1995) has used the term "piagetian" (with a small "p") to describe approaches to development that abandon specific content-laden hypotheses advanced by Piaget while preserving the essence of his ideas. The term is not a felicitous one, but the basic idea is one we like.

Chapter 6
Models and Maps

Suppose that you are visiting an unfamiliar city. Unless you have a great deal of time and energy, you will probably not explore the city simply by wandering around it. Instead, you might consult a map that indicates places of interest and shows where they are located. Or, you might ask someone who knows the city to tell you how to get to the places you want to see. Symbolic representations of spatial location, either in linguistic description or in various kinds of visual displays, serve to communicate information gained by one person to other people, saving them the necessity of personally exploring every area they visit. In this chapter we consider visual displays of spatial information, discussing what maps (and their close cousins, models) are, what is required for competence in using these symbolic tools, and what is known about children's development of the ability to understand and use models and maps. We defer to chapter 7 consideration of how spatial information can be communicated using language.

Several different theoretical positions have explicitly or implicitly guided research on the development of map understanding. One view is that children appreciate relatively little about maps until well into their elementary school years (Liben and Downs 1989, 1993). Liben and Downs link children's developing mapping abilities to their acquisition of skills described by Piaget as appearing slowly over the years of childhood, including use of projective systems of spatial representation, and understanding of proportionality.[1] A second perspective on the development of map understanding is to argue that it is innate (Landau 1986) or, at least, acquired quite early and apparently without any considerable exposure to maps (Blaut 1997; Blaut and Stea 1971, 1974). Nativist perspectives are often bolstered not only by studies of early competence but also by pointing out that the capacity to use and

---

1. More recently Liben (1999) has proposed a six-step developmental progression in which initial competence is achieved by preschoolers, but much remains to be achieved in later childhood and, indeed, by adults.

produce visually symbolic means of spatial representation appears to be a species-wide and cross-culturally universal competence dating back to prehistory (see references in Blaut 1991; Wilford 1981). A third way to analyze the development of map understanding, following Vygotsky, Bruner, Rogoff, and others, is to regard it as the cultural transmission of invented and socially shared symbol systems. In these views, supportive interactions between people more and less skilled in use of symbolic tools lead to acquisition of the tools by the less skilled. Studies of the development and use of navigation skills (e.g., Gladwin 1970; Hutchins 1995) fall clearly in this tradition.

These three positions differ radically in many ways, but nowhere more noticeably than in their treatment of input. Piagetians see input as the soil that nurtures the growth of children who follow their own timetable of development, nativists see input as a trigger to biologically determined programs, while Vygotskyans have popularized the metaphor of the scaffold that supports and instrumentally aids in development. Unfortunately, not much is known about the input children receive to aid them in acquiring map understanding. It is evident, however, even from informal observation, that exposure to maps is much less ubiquitous and much more individually variable than is exposure to our dominant symbol system, language. Liben and Downs (1989) suggest that this fact allows one to study the development of mapping skills as an example of the spontaneous development and construction of a representational system. However, exposure to maps is not always completely lacking in children's worlds, and thus, in looking at children's developing understanding, we are not examining a pure case of spontaneous invention. That is, the study of maps is not analogous to the study of "home sign" (e.g., Goldin-Meadow and Mylander 1984), where one can be confident that children were not exposed to sign languages at all. The limited but nonzero exposure of children to maps contrasts both with the study of normal language acquisition, where exposure is universal, and to the study of symbolic systems invented by children, where exposure is nonexistent. Sporadic and variable exposure leads to the likelihood that the sequences and age norms seen in studies of the development of mapping skills represent lower bounds on the developmental possibilities regarding what can be learned from the environment, while still not allowing for a window through which to examine spontaneous invention.

To preview, our reading of the literature on maps and models suggests that the field is currently ill placed to answer the questions about input that emerge from the theoretical controversies sketched above. There is a great need to examine more closely what children are taught about maps and what they can learn with different kinds and amounts of exposure to maps and other representational devices (see also Liben, in press). Such inquiry could also assess how tools such as maps and models can be used by children to acquire new knowledge about other domains, for instance, mathematics, geography, and science, and how using maps can transform children's understanding of spatial relations, for instance, by highlighting abstract spatial information (Liben and Downs, in press; Uttal 1999). Such matters are central both to understanding cognitive development and to understanding and improving education. For other spatial skills, such as mental rotation, it has been shown that practice and educational input lead to substantial improvements, both for initially low-ability and initially high-ability individuals (Baenninger and Newcombe 1989, 1995; Huttenlocher, Levine, and Vevea 1998), so there is reason to be optimistic about the effects of exposure to maps.

Despite the fact that we do not yet know what we need to know about spatial input, we do, however, have a solid basis for beginning to work on the problem of the relative roles of early starting points, developmental constraints, and cultural support in determining map competence. Recent work has led to considerable knowledge of developmental sequences of understanding and of rough ages of acquisition under current input conditions. Children begin to develop map competence early in the preschool years. They acquire an ability to link oblique and eye-level views of the same space by the age of 2 years (Rieser et al. 1982). They link iconically similar elements of differently scaled spaces by the age of 2.5 or 3 years (e.g., DeLoache 1987), and they can interpret more arbitrary symbols (e.g., blue rectangle as standing for blue couch) by 3 years (Dalke 1998). Indeed, by 3 years, they show understanding of arbitrary linkages such as "X marks the spot" (Dalke 1998). Shortly after 3 years, children show an ability to scale distance across simple spatial representations (Huttenlocher, Newcombe, and Vasilyeva 1999). These basic competencies are well ensconced by the age of 4 years, at which point children are set to tackle tasks that build on these competencies, such as mapping more complex spaces, establishing alignment between representations, and planning navigation. The path of such development is not, however, smooth. In fact even adults are not always adept at map use, and they display considerable individual variation in skill levels.

Maps and Models as Spatial Arrays

Before examining the development of understanding of maps and models, we need to examine what maps and models are. One way to define them is to explore how they are similar to and different from linguistic spatial descriptions and from each other, and to specify what is entailed in competent use of these symbolic systems.

Maps/Models versus Linguistic Description

Maps, models, and language can all communicate information about the spatial world. However, maps and models are uniquely suited to the representation of spatial information, with at least two virtues not shared by verbal descriptions. First, because maps and models exist in space, like the physical world they represent, they must, to be coherent, show the relations among represented locations in some metric fashion. Second, and again because maps and models exist in space, spatial relations among all symbolized locations must usually be displayed simultaneously.

To say that maps and models display metric relations is not to say that these relations are accurate. Often they are not, by accident, necessity, or design. Cartography is a matter of trade-offs and also of convention and cultural consensus. There is much debate in the field about how best to present information, and choices among various mapping possibilities are often made on the basis of a variety of aesthetic, graphic and functional considerations (Downs 1981, 1985). For instance, a certain element on a map (e.g., a subway station allowing for transfers to other lines) may be shown as larger than other elements because of its function and consequent importance to users, and not because of its physical size.

However, what we want to stress here is that the very need for mapmakers to decide such matters one way or another highlights the contrast between maps and language, since spatial information does not need to be metrically specified when speaking or writing. It need only be described categorically and thus at a very coarse grain (e.g., the store is "far from" or "near to" one's house). While the use of measurement language can supply more precise information (e.g., the store is "7.3 miles" from the house) or some specification of metric information can be made in phrases (e.g., the store is "beyond the playing field," or "about as far as the dry cleaner"), metric specificity is not required in speech; it is generally used only when exact information is considered essential. By contrast, any sensible map or model must necessarily supply metric information of some kind.

Maps and models also display information simultaneously. Because of this, relations not known by the map user can be read off as needed. Using a map instead of verbal directions to go somewhere is thus helpful if a detour turns out to be required or if one makes a mistake somewhere, since a new route can be devised by examining possible routes to the goal from the current location. In addition, referring to maps offers the possibility of avoiding the errors that commonly occur when people infer spatial relations from remembered information (Stevens and Coupe 1978). If you want to know which city is farther west, Reno or San Diego, or which city is farther north, Rome or New York, you will be much more likely to get the right answer if you consult a map or globe than if you rely on your own memory and inference abilities. Success is not certain; distortions due to the hierarchical nature of spatial coding can occur even when looking at simple maps (e.g., Acredolo and Boulter 1984). However, the competent map user has more of a fighting chance with than without a map.

Maps versus Models

Maps and models both represent the spatial relations among objects in the world simultaneously and in some metric fashion. As with other closely related categories, it is difficult to distinguish precisely between the two, although there is a clear contrast between prototypical instances (Blades and Spencer 1994). Typical models and typical maps differ along at least three dimensions. The typical model is three-dimensional and the typical map two-dimensional; the typical model symbolizes objects using small-scale replicas with a strong iconic resemblance to the objects they stand for whereas the typical map uses more arbitrary symbolic relations; the typical model generally stands for relatively smaller spaces than does the typical map, entailing a scale translation not as radical as that used in many maps.

Certain kinds of visual spatial symbolism are hard to classify, however, and exceptions and counter examples can be cited for each of the three generalizations mentioned above. With respect to two-versus three-dimensionality, there is the fact that relief maps are three-dimensional. With respect to the iconicity of symbols, consider that some maps use iconic symbols (e.g., a rectangle surmounted by a triangle to refer to a steepled church), and that even models have to strip away some real-world complexity (e.g., carving on furniture) in order to be able to symbolize. With respect to scale translation, note that the scale relations of a detailed U.S. Geological Survey map and of a display model of a large shopping center may differ very little. That familiar entity, a globe, shows clearly the difficulty of sharply differentiating between maps and models. A globe is three-dimensional (like a typical model), represents a very large-scale space (like a typical map), and is iconic if one views earth from the vantage point of a satellite but is noniconic for the earthbound. It has some aspects of both a model and a map.

What Does It Mean to Be a Map (or Model) User?

There are two kinds of information that may be coded from maps and models: element-to-element correspondence and correspondence of spatial relations (Bluestein and Acredolo 1979; Presson 1982). Alternatively, one may term these kinds of information representational correspondences and geometric correspondences (Downs 1985; Liben and Yekel 1996). Map competence involves using both of these kinds of correspondence to acquire information about the world, and more besides. Various reasoning and planning abilities are needed to use maps to solve navigational problems and to acquire new nonspatial information about the world. In addition complete competence involves the ability to produce as well as understand maps and models. In this section we briefly describe these components of competence.

Element-to-Element Correspondence

The most basic component in using maps and models is the ability to recognize element-to-element (or representational) correspondence. As we have seen, the symbolic relation in which a visual element on a map represents a real-world entity typically (although not inevitably) involves arbitrary relations between the two, as when a red circle represents a city. Symbolic relations in models are typically not arbitrary, but children still need to recognize that each object in a model has a one-to-one relation to a real-world object that it physically resembles (to a greater or lesser extent).

Research on element-to-element correspondence typically involves tasks in which children are asked to find a target object that is hidden under or behind a larger object. In such a task the location of the target is coded in terms of cue learning (see chapter 2). If children understand the relation of the model (or map) to the physical space, and remember the association of the target with a visible object, they should be able to find the target.

Correspondence of Spatial Relations

The second essential component in using maps or models is the ability to establish correspondence of spatial relations: the relations among elements in maps have a specifiable (although not necessarily simple) correspondence to spatial relations in the real world. Thus maps and models can be used to acquire distance and direction information about spatial relations among objects. Research on maps has dealt more extensively with correspondence of spatial relations than has research on models.

Two problems often arise in establishing spatial-relational (or geo- metric) correspondence. First, in order to use a map to learn about distances, viewers must appreciate the notion of *scale*. That is, they must understand that most maps evenly transform real-world dis- tances into smaller ones and that this transformation rule can be used to judge real-world distances. Although such transformation is ideally done with mathematical precision, no map can be interpreted without at least rough or intuitive scaling. For instance, in looking at a treasure map, one needs to learn that the treasure chest is buried ten paces from the tall pine tree and twenty paces from Skull Rock. If one ignores scaling and takes away the information that the chest is two inches from the tree and four inches from Skull Rock, the treasure hunt will be doomed.

Second, competent map use includes an understanding of *alignment*, that is, knowledge of the relation of the self both to the physical world and to the represented world. Maps can be used for certain purposes without such understanding (e.g., to find out how many churches exist in a given town, or to discover that Montreal is north and east of Toronto). But without an understanding of alignment, maps cannot be used for navigation. Establishing alignment is simplest when the map and the viewer's vantage point on the world are similar in elevation (e.g., as they may be for the layout of small objects in a room). Even then, viewers must ensure that their representations of the map and the world are in alignment in the plane, either by physically turning the map or by mentally rotating one or the other representation. Estab- lishing alignment is even more difficult when a map gives an aerial view of a space to which a viewer does not normally have such access, as do most maps of large-scale spaces.

Using Maps
Being able to work with correspondences of elements and correspon- dences of spatial relations is not sufficient to ensure competent use of maps for navigation or for gaining knowledge about nonspatial aspects of the world. To navigate using a map, the map user needs additional capabilities, including the ability to systematically locate starting and ending points, to trace physically possible routes between two points, and to compare alternative routes in terms of distance as well as other variables related to transportation efficiency (e.g., ruggedness, if walk- ing, or traffic conditions, if driving). To gain knowledge about un- known areas from maps may require a great deal of nonspatial knowledge. For instance, a geography student can use knowledge of

latitude and altitude to identify crops that might potentially grow in an area under study.

Making Maps

Map competence also involves the ability to construct as well as to use a map. Constructing a map is different from, and more demanding than, using a map. The mapmaker must depict many different spatial relations simultaneously, not all of which may be directly known. Therefore, making maps can involve working out unknown spatial relation by inference, or by making new observations of the world. Really accurate mapmaking requires the use of measurement tools and surveying techniques, and the solution of difficult mathematical and graphic problems. Indeed, cartography is a complex subject in which there is ongoing research and controversy (see Downs 1985).

Summary

Maps and models, unlike linguistic description, exist in space and therefore obligatorily communicate both information about spatial elements and their relations. Such information must be integrated and simultaneously available. Map-and-model competence involves the mastery of element-to-element and geometric correspondence, across variations in scale and viewpoint, as well as the ability to reason about and construct such representations.

Developmental Studies of the Components of Map and Model Use

There are developmental studies of some, but not all, of the components of competence with maps and models. In this section we review the available information on several issues. We begin with understanding of element-to-element correspondence, first in models, where the relation is generally more iconic, and then in maps, where the relation is often (although not always) arbitrary. We go on to consider spatial-relational (or geometric) correspondence, including understanding of scale and of alignment. We briefly discuss using maps for navigation and to acquire geographic knowledge, and constructing one's own maps. Finally we consider Case and Okamoto's (1996) approach to spatial-symbolic tasks, including map tasks but also dealing with copying and drawing tasks.

We note, in advance, that much of the research to be reviewed has an age-normative, descriptive cast. That is, researchers have typically defined some aspect of map competence and asked at what age children evidence mastery of it. While such questions are a sensible start for a research program, we must remember, as argued also by Liben

(1999), that describing age norms under current conditions is only a prelude to asking how development works, and how it might work under different conditions. With different amounts and kinds of input, children might be able to understand maps earlier and progress more quickly to higher levels of competence. (With other kinds of input, they might of course be delayed and ultimately less competent.)

Element-to-Element Correspondence

Models
While a representation such as a dollhouse seems, to the adult eye, an almost transparent symbol (i.e., a miniature couch seems to stand for a real couch in an obviously compelling way), very young children often do not appreciate the symbolic relation between models and real-world spaces (DeLoache 1987, 1995a, 1995b). Indeed, children's reactions to prototypical models turn out to raise interesting issues about the acquisition of spatial symbolism.

DeLoache (1987) first reported the phenomenon of rapid developmental change in the use of models in a study of children's ability to find hidden objects. Working with a lab room equipped as a typical living room and a model of that space, she found that children as old as 2.5 years did not seem to appreciate that if they saw an object hidden under the couch in the model, they could use that information to find the larger version of that object, hidden under the couch in the living room. (Vice versa, they also did not appreciate that seeing an object hidden under the large couch implied the location of a smaller version of the hidden object in the model. Henceforth, for simplicity, we speak only of learning about the room by consulting the model.)

The data from DeLoache (1987, 1989) are shown in figure 6.1a. All children did quite well in retrieval 2, in which they were asked to find the object hidden in the model by looking in the model (or the one in the room by looking in the room). Their good performance in this condition shows that the difficulty the younger children had with finding toys in an analogous location (retrieval 1) is not due to difficulty with location memory itself. The data show a startling increase between 2.5 and 3 years in the ability to appreciate the relation between elements in the model and in the larger space. Figure 6.1b shows that this variation is linked, within age, to whether or not children show evidence of understanding the basic room-model correspondence.

The probability of children linking model objects to their larger counterparts is not all or none, however, appearing as a general shift toward the age of 36 months. Performance is affected by the degree of

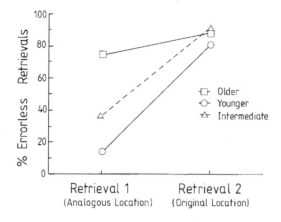

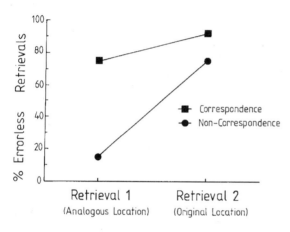

Figure 6.1
Top panel shows percentage errorless retrievals for analogous locations (room-model correspondences) and for original locations for older children (36 months), younger children (30 months), and intermediate children (33 months). Bottom panel shows percentage errorless retrievals for intermediate children divided by whether or not they showed independent evidence of understanding room-model correspondence. From DeLoache (1989).

iconic resemblance between the objects in the model and the objects represented. For instance, even 2.5-year-olds can succeed at the task when the scale of the model and scale of the real room are very similar, while even 3-year-olds have difficulty with the task when the objects vary in appearance, for instance, when their colors are changed (De-Loache, Kolstad, and Anderson 1991).

DeLoache and her associates have explored these phenomena in a series of studies, arguing that the essence of children's early problem with using models is in seeing the model as a symbol at all—not in, for instance, a basic capacity such as appreciating the perceptual similarity between the symbol and the referent. The children can readily point out correspondences between the model and the room; the question is whether they will take one as imparting information about the other (DeLoache 1995a). The problem with models, DeLoache suggests, is essentially that they are too real. They are real three-dimensional objects in and of themselves. Therefore children see them, for all their similarity to real-world objects, as objects in themselves rather than as symbols standing for something else. They have difficulty with dual representation of the models as both objects and symbols.

There are several lines of evidence for this conclusion. One kind of evidence comes from the fact that children of 2.5 years have greater success in interpreting pictures as guides to locating objects in the real world than they have with models (DeLoache 1991; Dow and Pick 1992). At first blush, this seems odd, since pictures, even photographs, seem to an adult to represent the world less well than a model does: pictures are two-dimensional, with certain objects and vistas possibly obscured, and they may lack or distort color, or delete a good deal of visual and textural detail. But DeLoache argues that this is just the point: a picture is obviously not an object with which one can interact or play but rather an object whose function is to stand for something else. Interestingly, experience performing the location task with pictures actually leads to improvement in children's performance on model tasks (DeLoache 1991). DeLoache argues that experience with pictures facilitates the "representational insight" that models (like pictures) are meant to stand for something. Seeing pictures as symbols is not itself inevitable but appears to simply become available earlier than seeing models as symbols; at the age of 2 years, children have difficulty conducting searches based on information from pictures (DeLoache and Burns 1994).

A second line of evidence for the representational insight hypothesis comes from examining the effects of physical interaction with models. Most observers would suggest, following Piaget and probably also common sense, that the more children interact with a medium, the more they will understand it and be able to use it. But the representational-insight hypothesis suggests that treating a model as a plaything works against treating it as a symbol. In line with this approach, using the model as a plaything turns out to hurt performance in using the model to find hidden objects, while not interacting with the model actually enhances the potential for children to see the model as a symbol or a tool (DeLoache 1995b).

A third line of evidence comes from another way that children can achieve representational insight earlier than usual. If children can be persuaded that the model and the real-world space are one and the same thing, their likelihood of treating the model as informational increases because, in a way, representation is not involved. DeLoache et al. (1997) told children they had invented a "shrinking machine" that allowed for a larger space to be miniaturized. After witnessing an odd-looking machine emitting strenuous noises apparently reduce a larger model to a smaller one, children were able to find an object hidden in the larger model in the "miniaturized" version. In this case children need not achieve "representational insight;" if children believe the cover story, they need not appreciate the model as a representation at all.

In summary, DeLoache's research has shown us that treating a model as a symbol is not automatic for children who are 2 or 3 years old. Within this age range, the task may be made easier or harder by varying factors such as iconic resemblance, scale, or opportunity to interact with the model. But the main message is that the idea that a model represents something is a fragile one whose use requires initial support. At some point, sometime after 3 years or by age 4 at the latest, children are in comfortable possession of the idea that models can be tools for thought.

Maps
Since one of the differences between typical maps and typical models is in iconicity of representation, a basic question about understanding of element-to-element correspondence in maps is whether children have additional difficulty in interpreting the arbitrary symbolic relations often used on maps. Presson (1982) found that 6-year-olds had no difficulty interpreting an X as representing the location of a ball they were to find; similarly Sandberg and Huttenlocher (1997) found that 6-year-olds had no difficulty using a blue dot to represent their starting point and a gold star to represent their destination in a navigational task. Whether children younger than 6 years succeed with arbitrary pairings is more controversial. Liben and Yekel (1996) report that 4- and 5-year-olds say that even somewhat iconic representations in maps

(e.g., rectangles for tables) are deficient (e.g., "they don't have legs"). However, the same children also spontaneously recognized a plane view of their classroom using such symbols as a representation of their classroom, so their comments may not indicate a basic difficulty in establishing correspondences, simply an aesthetic judgment. Indeed, Dalke (1998) recently reported that even young 3-year-olds can recognize the correspondence between somewhat iconic representations and their real-world counterparts, and that the same children are above chance in interpreting arbitrary relations such as "X marks the spot," although older 3-year-olds and 4-year-olds are better.

Other map studies examining children's ability to interpret elements with a good deal of physical resemblance to the objects for which they stand substantiate, as one would already expect from our discussion of the work with models, that children understand such symbolization by the age of 4 years. For example, Bluestein and Acredolo (1979) found that children as young as 4 years appear able to understand element-to-element correspondence when elements on maps are schematic renditions of objects in the real world (i.e., a drawing of an elephant representing a toy elephant). Uttal (1994, 1996) found that 4-year-olds could learn locations from a map consisting of small photographs of target objects. Landau (1986) reported that a blind child, at the age of 4, could interpret small objects explored with her fingers as representing larger entities in a room.

Work on children's ability to interpret aerial photographs, including their ability to recognize plane views of environmental features generally seen at eye level, addresses another aspect of element-to-element correspondence. Blaut, McCreary, and Blaut (1970) reported informal evidence that children as young as 5 years can interpret some information in aerial photographs; for instance, they were said to recognize houses and roads in these photographs. Blaut et al.'s finding is consistent with the abilities at element-to-element correspondence shown by 4-year-olds (Bluestein and Acredolo 1979; Landau 1986; Uttal 1994, 1996). The finding indicates, in addition, that at least by 5 years, the degree of iconic resemblance can be reduced by radical changes in viewpoint and scale and that the correspondence can still be discerned.

However, interpreting aerial photographs may not be as simple as Blaut et al.'s observations would lead one to believe. Liben and Downs (1989) have observed instances in which children viewing aerial photographs have insisted that roads cannot be represented by the lines they see on the photographs because "they're wider than that" or in which children have interpreted a round orange water storage tank seen in aerial view as a piece of fruit, despite the fact that a much smaller rectangle next to it has just been called a house. When object depictions are unfamiliar, as when seen from above, discerning what they are may be difficult for elementary school children (and perhaps, in certain circumstances, for adults as well). One aspect of the problem may be that consistent interpretation of aerial photos requires consistent use of scale information (an issue to be discussed shortly).

Summary
Children begin to acquire various kinds of symbols in the second year of life, as they learn words and begin to show evidence of symbolic play. Acquisition of the ability to interpret graphic elements as existing in a "stands for" relation is apparently slowed by the fact that pictures, and, yet more obviously, small objects, are things in themselves. However, when the realization of a relationship comes, it apparently comes swiftly in the form of an insight. The timing of the insight varies as a function of many factors, but it is largely secure by sometime after 3 years.

The lack of investigation of early understanding of arbitrary symbol relations in maps is unfortunate. The fact that arbitrary symbol-referent pairings, as occur in language, can be learned much earlier suggests that perhaps with some appropriate guidance, children could appreciate such pairings from a fairly early age. The work of Dalke (1998) supports the idea, but more research is needed. It is possible that learning arbitrary symbol pairings might be easier than interpreting pictures or models (which are things in themselves), or aerial photographs (which require coordination of scale and recognition of objects from unusual viewpoints).

Spatial Relations in Maps and Models

As we discussed in chapter 2, the location of a smaller object can sometimes be coded as juxtaposed with or hidden by a larger one (e.g., a stuffed animal under a pillow or behind a chair). Understanding element-to-element correspondence in maps or models is sufficient to find objects in the world using these representations if the objects are hidden in, on, or immediately next to a represented element. However, such cue learning works to locate objects only in very particular circumstances, and does not involve use of relations of distance and direction (i.e., place learning). A much more powerful use of maps and models is to indicate the coordinated spatial relations among objects, using distance and direction estimates. There are several sources of evidence on the development of use of maps and models to indicate spatial relations and to allow for acquisition of distance and direction information across changes in scale and orientation.

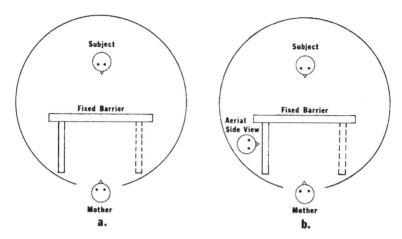

Figure 6.2
Top view of the procedures in Rieser et al. (1982).

One Early Starting Point

A very simple situation in which to study the coding of distance across a change in viewpoint is shown in figure 6.2. Rieser et al. (1982) showed infants an aerial view of their mother behind a barrier with one closed side and one open side. The infant was then placed back on the floor and the mother called for the baby to come to her. At 25 months (but not at 21 months), infants systematically headed for the open side on the basis of their aerial view, indicating an ability to learn about eye-level space from an oblique view in a situation where no scaling information was required. They could do this even when they had been given a side view of the apparatus, which differed from their perspective when they were set down facing the barrier. This study shows two interesting things about the early basis of mapping ability: both how early some prerequisites may be available, and that even these simple prerequisites may take some time to be constructed, appearing between 21 and 25 months.

Scale

From the Piagetian perspective, elementary school children should have difficulty dealing with scale, since understanding scale depends on understanding proportionality, said to be a formal-operational ability not appearing until early adolescence. Children actually do show fair amounts of inaccuracy in interpreting maps to give distance and direction information, as we will see shortly, but the their problems may not derive from a fundamental inability to comprehend scale.

Huttenlocher, Newcombe, and Vasilyeva (1999) recently studied 3- and 4-year-old children's use of a map of a five-foot-long sandbox to acquire information about the location of a toy hidden in the sandbox. The children were shown a dot in a small rectangle drawn on a piece of paper and asked to use this information to find a disc that had been hidden in the sandbox. To do this successfully, the children had to be able to encode proportional distance. In fact the 4-year-olds, and even a good number of the 3-year-olds (15 of the 25 tested), were able to perform very well on this task, showing errors which averaged only 4 inches. This performance indicates an ability to scale observed distances in a simple situation: one in which only one object had to be considered at a time, in which only one dimension had to be considered, and in which the map was physically available to guide search.

Interestingly a certain number of the 3-year-olds (10 of the 25 tested) were unable to use the map at all, with 6 of the 10 children simply pointing repeatedly to the same place, indicating a lack of understanding of the task. The fact that the 3-year-old group could be easily dichotomized into those children who essentially did not understand the use of the map to give distance information, and those children who did as well as the 4-year-olds, suggests that what we may be seeing here is an insight similar to the representational insight studied by DeLoache.

While children can coordinate distance information across scale by around the age of 3 or 4 years in these studies, other work seems potentially contradictory, showing older children having trouble on scaling tasks. When asked to reproduce configurations of objects learned from a map, 4- and 5-year-olds reconstruct angular relations correctly, but they are not very accurate on scale, showing both shrinkage and expansion relative to a correct layout (Uttal 1996). Preschoolers are better at scale translation when the configurations of objects were symmetric (Uttal 1996), but of course symmetry is uncommon on real maps. Preschoolers also have difficulty with metric placement of items, one at a time, in familiar classrooms (Liben and Yekel 1996). Children of 6 and 7 years do better on reproduction tasks than preschoolers, although not as well as adults—who were nevertheless not perfect translators of scale (Uttal 1996). Children as old as 7 years make substantial errors, usually of underestimation, when asked to use a map to place objects in a real space, and children through the age of 10 years perform worse than children of 11 to 13 years (Wallace and Veek 1995).

Why do older children have these problems with scaling when studies of younger children show some competence in scaling? The answer may lie in a combination of changes in task understanding, increased ability to deal with task demands, and suppression of irrelevant task strategies. First, preschool children's scaling ability may depend on relative or proportional coding (e.g., "a quarter of the way across") rather than on the construction of mathematical ratios. Use of such ratios to calculate absolute distances may be a much later achievement. Second, older children show difficulty in scaling in situations where there are many spatial relations to consider and hence considerable demands on working memory to maintain a constant scaling relation over these relations. Ability to deal with working-memory demands may well improve developmentally. Third, young children may also be hindered in scaling tasks irrelevant response preferences, for instance regarding placing items in what look like "empty spaces" on maps (Liben and Yekel 1996). Such strategies may well evolve to more adaptive preferences with age and experience.

The Piagetian argument that use of scale depends on acquisition of proportionality at the stage of formal operations requires some reconsideration. First, children's ability to deal with scaling is present in quite young children in simple situations and at least roughly present in more complex ones. This ability may depend on a relative scaling rather than computation of absolute distances, but the whole issue of when and whether absolute equivalencies are computed requires further research. Second, the ability to deal with scaling shows more evidence of an initial early insight followed by steady and progressive quantitative improvement than evidence of abrupt change late in childhood or in early adolescence. Study of the advent of mathematical computation in scaling might show such a late transition, but the relevant work has not been done.

A Puzzle: Spatial Relations in the Model-Room Task
There is evidence from DeLoache's model research that young children use spatial relations in judging the overall similarity of model to room. Marzolf, DeLoache, and Kolstad (1997; see also Marzolf and DeLoache 1997) found that 3-year-olds were more likely to recognize the model-room relation if objects occupied the same relative positions in the model and in the room, and less likely to see the relation if spatial positions were different.

However, another aspect of the Marzolf et al. study is puzzling. When children saw a room and a model with objects arranged differently in the two spaces, they could have searched for the hidden object using a contiguous cue (e.g., "under the pillow") or in terms of distance and direction, from either the self (e.g., "back right corner of the display") or the external framework (e.g., "closest to the door of the testing room"). These three possibilities were placed in conflict by the furniture being arranged differently in the room and the model.

The data showed that the children relied on the contiguous cues (i.e., they looked for the hidden object "under the pillow") rather than position in the room/model (i.e., "back right corner" or "closest to the door"). One possible conclusion is that 3-year-olds cannot use distance information from models, in contradiction to the Huttenlocher et al. (1999) results.

Fortunately the data do not compel this conclusion, and another interpretation of the children's behavior seems more plausible. When children are presented with a conflict between different spatial coding systems (each quite reasonable in its own way), their performance may speak only to their preferences and guesses about the experimenter's preferences, rather than to their abilities. They may prefer, at least in situations where they don't know what coding system will turn out to be useful, to rely on cue learning. But such a choice may indicate a hunch in the face of uncertainty, not ability or inability. Reliance on cue learning may actually be basically sensible when there is uncertainty. After all, if we move a couch, an object hidden under a pillow on it is quite likely to go along.

Alignment

When maps are aligned, physically or mentally, with the world, it is possible to straightforwardly interpret directional information gained from the map. That is, "right" on the map means "right" in the space, and "front" means "front," and so on. When maps are not aligned with the world, bringing them into alignment to answer directional questions (e.g., "which way to Smithtown?") requires either physical rotation of the map, or mental rotation of the map information.

Physical rotation of a misaligned map or model into alignment with the world is probably a fairly easy task. Children as young as 3 or 4 years can certainly physically rotate a model to answer questions about another's perspective (e.g., Borke 1975; Rosser 1983), likely by focusing on an individual item in the array (Newcombe 1989). Physically rotating a map or model to congruence with one's own perspective seems analogous: children could note what they are looking at (e.g., a particular window) and move a map of the room until the depiction of the window was in front of them. Thus, if children understand the necessity for alignment and are allowed to physically move a map (a method of dealing with misaligned maps recommended in the *Boy Scout Handbook*), they should be able to succeed by 5 years at least. Both Uttal, and Sandberg and Huttenlocher, have observed that it is difficult in fact to stop children of this age from physical rotation of a map.

By contrast with physical rotation, mental rotation of misaligned maps appears to be very difficult. That is, many people, including

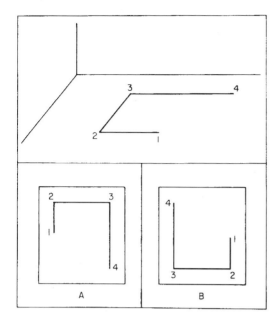

Figure 6.3
Four-point paths used by Levine et al. (1982) (top panel) and two maps of that path
(bottom panel). If a person has represented the path in an orientation-specific way with
point 1 at the left, map A is aligned and map B is not.

many adults, struggle to answer directional questions about locations
when map information is misaligned with the direction they are facing
(Levine, Jankovic, and Palij 1982). Levine et al. asked college students
to learn paths of the sort shown in figure 6.3. Half the subjects learned
the map in one orientation and half learned it in the other orientation.
The students were then asked to imagine being at one point on the
map, facing another point (e.g., at point 4 facing point 3) and to point
to another location (e.g., point 1). This task was much easier when the
map (or mental image of the map) was aligned with the to-be-imagined
space than when it was not. In the latter case, Levine found that adults
merely guessed as much as half the time.

Children have been found to have a good deal of difficulty, decreas-
ing with age, with various tasks which involve dealing with misaligned
maps and models (Blades 1991; Bluestein and Acredolo 1979; Laurendeau and Pinard 1970; Liben and Downs 1989; Piaget and Inhelder
1948/1967; Presson 1982; Pufall 1975; Pufall and Shaw 1973). According
to Piaget, children's problems with misaligned maps and models stem
from their reliance on topological spatial coding and their lack of
understanding of projective spatial relations. An alternative view,

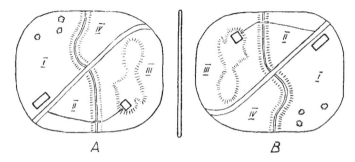

Figure 6.4
Model village used by Piaget and Inhelder (1948/1967). In model A, a stream runs from top to bottom. On the right is a hill topped by a house with a yellow roof. A road runs diagonally, with a house with a red roof on its left. There are three trees around a small hill in the upper left. Model B is the same as model A, except that is is rotated 180 degrees.

already presented in our discussion of perspective taking in chapter 5, is that problems with misaligned maps and models stem from the fact that dealing with them involves dealing with conflicting frames of reference.

Piaget and Inhelder (1948/1967) began the study of children's problems with misaligned arrays. They examined children's ability to copy an arrangement of buildings and animals from one model to another not in alignment. Both models were marked with a road, a railroad track, and various other features (see figure 6.4). Children had great difficulty reproducing the location of figures, especially when distance and direction from landmarks on the model had to be considered (i.e., when the to-be-placed object could not be located directly on or beside a landmark). Similar findings were reported by Laurendeau and Pinard (1970), Pufall and Shaw (1973), and Pufall (1975). The traditional interpretation of these data is that children can only code location associatively (or topologically). Another possibility, however, is that the children's performance is disrupted because placing objects correctly with respect to the frame of reference of the model features requires placing objects incorrectly with respect to the frame of reference of the wider room. Indeed, as we discuss in chapter 7, there are cultures in which placing objects with respect to an absolute frame of reference, such as that of the room, is considered the correct response. The existence of cultural variation in use of frames of reference undermines the idea that these tasks index a necessary developmental sequence from topological to Euclidean coding.

Children also have difficulties dealing with misaligned spaces in studies asking them to find a single hidden object whose location is

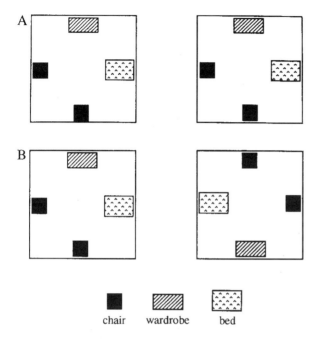

chair   wardrobe   bed

Figure 6.5
Pairs of model room layouts used by Blades (1991). In A, rooms are aligned; in B, rooms are rotated with respect to each other. Each room contains two identical chairs and two unique items. From Blades (1994).

depicted in a model or map, although developmental improvement seems to come earlier in these studies than in the Piaget-Inhelder copying task. Blades (1991) used two models of equal size, each of which contained four objects under which a target might be hidden: two distinctive objects and two identical ones (see figure 6.5). The models were sometimes aligned with each other and sometimes not, and the layouts were sometimes fairly symmetrical and sometimes arranged so as to place one of the identical objects next to a distinctive object which could serve as a landmark. Finding a target under a distinctive object requires only cue learning, and as would be expected from prior research, even 3-year-olds could do this. However, 3- and 4-year-olds failed in the nonaligned condition when targets were hidden under nondistinctive objects; 5-year-olds were successful. Again, the data have a straightforward interpretation under a conflicting-frames-of-reference hypothesis; children may be using other frames of reference than the internal relations among the model elements. In particular, they may be relying on the relation of the model elements either to the room cues or their own body, two primary means of coding spatial location.

Other studies have used maps misaligned with the rooms they represent, rather than two models. Bluestein and Acredolo (1979) found that 4-year-olds were unable to use a map rotated 180 degrees to find a toy elephant, making a large number of egocentric errors. Difficulty with rotated maps was especially evident when children were reading the map inside the room, although it was also significantly present when the map was outside the room. This advantage for reading maps when outside the room provide support for a conflicting-frames-of-reference hypothesis. When children are not in the room, they do not encode elements on the map with respect to the external landmarks in the room (e.g., door, windows) and are more likely to encode the relations among the mapped elements, as needed to locate a target.

Presson (1982) found that problems with 180 degree misalignments of maps may persist to 8 years. He also found that 90 degree misalignments were easier than 180 degree ones, for both 6- and 8-year-olds. The fact that left–right and front–back relations are systematically reversed by 180 degree misalignments, but not in other kinds of misalignment, may account for this difference. Liben and Downs (1993), like Presson (1982), found that children's difficulties with indicating positions on misaligned maps persist through the early elementary school years. Because children in their study remained in the space in which they originally learned the spatial relations, the difficulties observed can as easily be expressed in terms of conflicting frames of reference as using Liben and Downs's preferred terminology of understanding of projective spatial representation.

Summary
Many studies show that children have trouble dealing with unaligned spaces. The ages at which they begin to be successful vary from 5 years (Blades 1991; Bluestein and Acredolo 1979) to preadolescence (Laurendeau and Pinard 1970; Liben and Downs 1993; Piaget and Inhelder 1948/1967; Presson 1982). On some tasks many adults continue to have marked difficulty (Levine et al. 1982). We believe that the fact that alignment is one of the more difficult aspects of map use, even for adults, derives from the conflict that unaligned spaces inevitably create between frames of reference and, often, the need to correct for misalignment by performing mental rotation. Various strategies can be used to deal with conflicting frames of reference (Newcombe and Huttenlocher 1992), and these may be discovered and used with greater consistency as children grow older. An additional source of developmental improvement may be increases in the efficiency of mental rotation (Kail 1988). In support of this argument, Scholnick, Fein, and Campbell (1990) found correlations between success in map use and mental

rotation ability in 4- to 6-year-olds. Increases in mental rotation ability may themselves be explained by various factors, among which practice and educational input are important (Baenninger and Newcombe 1989, 1995; Huttenlocher et al. 1998).

Using Maps and Models

Our discussion so far has concentrated on children's ability to use maps to acquire very simple spatial facts about the world, basically the location of individual objects or small sets of objects. Maps are sometimes used in this way (e.g., a simple treasure map), but more commonly, people use maps to plan routes and to find their way, or to represent and learn about linkages of nonspatial and spatial information (e.g., maps showing population density). In this section we turn to examine what is known about children's ability to use maps to accomplish tasks of this sort.

Navigation

A basic question about using maps for navigation is whether viewing maps allows people to learn routes more easily than would people who have not been able to see a map, or allows them to anticipate spatial relations they will encounter when navigating. In fact Uttal and Wellman (1989) and Scholnick, Fein, and Campbell (1990) have found that children as young as 4 years benefit from maps and can use them to guide navigation in a simple situation. This conclusion, however, must be considered in the context of the nature of the maps used: the maps had low levels of symbolic complexity and made no demands on alignment or scaling. In addition children were simply shown routes on the maps; they were not required to plan or compare routes.

The maps Uttal and Wellman used are shown in figure 6.6. Interpreting the symbols was not difficult, since small photographs of stuffed animals stood for the larger animals located in each "room." The maps were shown to children before they explored the space, in alignment with the space. The relations among the animals were quite regular, following a grid pattern, so categorical relations such as "in front" or "next" were sufficient to encode both the map and the real space; distance and direction did not need to be precisely calibrated from map to world. Finally the dependent measures were how quickly children learned a single U-shaped route (as compared with map-naive subjects) and how much map experience improved children's ability to point to unseen toys.

Planning routes is clearly more demanding than simply following routes on maps already indicated by someone else. Fabricius (1988)

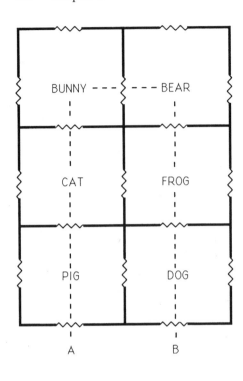

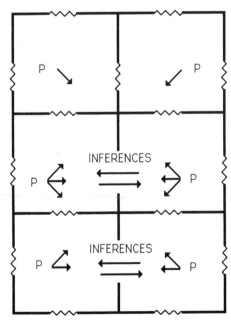

showed that 5-year-olds perform considerably better than 4-year-olds at planning a search of a small three-location array. Sandberg and Huttenlocher (1997) showed that by six years, children could plan routes through a complex environment in which many alternatives were possible. In addition the children could use distance information on the map both in this planning and to stop in good proximity of their target, located along a lengthy hallway. However, 6-year-old children may still have something to learn about route planning. They do as well as older children at following a route when a path is indicated by the experimenter but do less well at planning efficient routes (Cramer and Presson 1995).

Both Sandberg and Huttenlocher, and Cramer and Presson, used maps of fairly regular rectilinear spaces. People would likely find it more difficult to plan routes on more naturalistic maps, or with various circuitous routes. Experience suggests that even adults have difficulty in planning routes in such circumstances, especially when distance is not the only variable but factors such as road quality, terrain, and traffic density need to be considered as well. Bailenson, Shum, and Uttal (1997) have recently shown that college students use heuristics to plan routes, for instance, preferring to leave their starting point on a long straight route segment. Thus they may pick different routes for getting to point A from point B than they pick from getting to point B from point A.

Acquiring Geographic Knowledge
Maps can be used for other purposes than planning or following routes. They also serve as data displays to clarify and summarize information regarding population, language communities, climate, or the like. They function as tools in teaching and learning facts about the world, such as the names of capitals or the relative location of countries establishing trade relations or ties of sovereignty. They can display the structure of events unfolding over time, such as the sequence of moves made over three days by the troops of the Union and the Confederacy at the Battle of Gettysburg. Visual presentations of this sort often communicate knowledge more succinctly than equivalent verbal exposition, and they can make vivid the significance of certain statements, such as the claim

Figure 6.6
Top panel shows the layout of the playhouse used in Uttal and Wellman (1989) and the U-shaped route. The bottom panel shows the positions and directions used in the inferencing and pointing measures of learning. P is the position from which child was asked to point, arrows represent direction of points, and arrow-labeled inferences indicate directions of the four inference questions.

that had the Confederacy gained control of Little Round Top on day 2 of the Battle of Gettysburg, they would likely have won the battle. In fact, making or looking at various kinds of spatial representations can be a tool for achieving new discoveries and insights, not just communicating existing ones (Liben, in press). Unfortunately, although maps are used with some frequency in newspapers (Monmonnier 1989) and in schools, very little is known about how and how well people use them or how they can be educated to use them to best advantage. The low levels of geographic knowledge displayed by the American population in several surveys suggest, however, that such research might be of practical importance.

Construction of Maps and Models

Because children's graphic abilities are usually not adequate for them to draw maps, investigators have generally examined construction of a model for which small-scale replicas of to-be-placed items can be made available, or they have given children drawings or photographs of objects to position on a piece of paper. Children can move these around as they attempt to make a satisfactory model; experimentation in the course of drawing, even with an eraser handy, may be more difficult.

An initial study by Siegel and Schadler (1977) suggested that children, at least by 5 years of age, have few difficulties constructing models of a familiar space. Most 5-year-olds in the Siegel and Schadler study performed well in constructing a model of their kindergarten classroom. Later studies, however, seemed to indicate more difficulties with model construction. Siegel et al. (1979) had 5-year-olds learn the location of eight buildings constituting a room-size model town. They found that 5-year-olds had difficulty making a model of the town. Children's problems in the Siegel et al. study might stem from inadequate learning and memory for the layout, given that they had equivalent difficulty recreating the layout of the town using the original buildings. However, Liben, Moore, and Golbeck (1982) argued that memory for the location of objects is not the whole of the problem. Liben et al. found that even when children as old as 5 years are modelling familiar spaces (again, their classroom) with visual access to the space, they are only able to place roughly half the items correctly (see also Golbeck, Rand, and Soundy 1986).

Before accepting that children through the age of 5 years have difficulty constructing models (or even replicas), it is important to consider an alternative explanation for the findings. The analyses in these studies used a very specific kind of dependent variable, namely accuracy of placement of individual items relative to that item's correct location. That is, item placements were counted as correct if they were within a specified distance of the correct location, or errors were analyzed (i.e., distance of a placed object from the correct location). As Huttenlocher and Newcombe (1984) and Naveh-Benjamin (1987) have pointed out, however, accuracy as assessed by measures such as these may be very low, even when configural accuracy is high. That is, if configurations are rotated, or expanded or contracted, the placement of objects relative to each other may be quite sensible, even when very few of the objects are close to their correct locations. Uttal's (1994, 1996) studies substantiate in detail that when children are asked to reproduce sets of locations, they often get the scale wrong but the relative locations right.

Good performance in relative-location terms but poor absolute performance is, we think, better thought of as a difficulty with scaling distance than as a global difficulty in coding and reproducing location. Because studies of model construction have not looked at configural accuracy, we do not know whether it is present in children's models. A few individuals may show thoroughly scrambled placements (e.g., Siegel and Schadler discuss a girl called Buffy whose models of her classroom were unrecognizable), but such examples do not demonstrate that such highly inadequate performance is common.

There are other factors besides scaling difficulty that can make constructing a model or map more difficult than using one. First, in making a model or map, children have to create many different spatial relations, whereas in a particular test of map use (e.g., finding a toy elephant), they only need to code a small number of spatial relations (perhaps only one). The need to create many different spatial relations also leads to a second problem, the issue of sequential dependency. If early placements are off (e.g., if the child makes a mistake in placing the teacher's desk when reproducing a classroom), then placement of subsequent items which may have been coded with respect to that early placed item (e.g., a wall map located behind the desk) will inevitably be erroneous relative to the correct location (even if accurate relative to the landmark used). Worse, if early placements are off, the whole task may simply become globally more confusing. Third, map construction is likely to be harder than map use simply because children have to perform the task in order to achieve an abstract goal (i.e., constructing a representation when asked by an adult) rather than simply working to accomplish a desired goal (e.g., finding a toy).

In short, children may find it difficult to construct models, but we do not yet know how severe or fundamental their problems are. The initial impulse of many investigators who see jumbled models is to infer that they reflect poorly structured and even incoherent spatial memories. But an alternative possibility is that inaccurate models may coexist with basically accurate spatial memories that nevertheless contain a greater degree of inexactness and uncertainty than is typical for adults, and that children have additional difficulties making models because of the necessity to maintain scaling relations over multiple spatial relations, to deal with initially incorrect placements, and to maintain interest in a lengthy and abstract activity. We know even less about children's ability to construct (or to learn to construct) maps, that is, to use arbitrary symbols to represent large-scale spatial relations.

Mapping as an Instance of Spatial-Symbolic Skill

Case and Okamoto (1996) have recently proposed an account of the development of spatial-symbolic skill as part of a domain-general account of cognitive development, focusing on what they call central conceptual structures. They have applied this framework to other domains as well as space, including numerical understanding and narrative construction. They argue that children's thought progresses from a predimensional period, at about 4 years of age, in which understanding is purely qualitative, to a unidimensional period, at about 6 years of age, in which relations can be constructed quantitatively along one dimension at a time, to a bidimensional period, at about 8 years of age, at which relations can be constructed along two dimensions, to an integrated bidimensional period at age 10, in which relations along two dimensions can be related to each other.

To investigate the application of this framework to the spatial domain, Case and Okamoto investigated children's performance on tasks such as seriation, copying arrangements of checkers, picture drawing, map drawing, and map following. These tasks were selected because they could all be analyzed in terms of children's utilization of mental reference lines (most notably, a ground line). A predimensional performance is then one in which objects are placed or drawn without regard to any mental reference line; unidimensional performance involves using such a line; bidimensional performance involves using two referents, often a ground line and an orthogonal vertical line; and the last stage involves integration of the two reference lines. For the map drawing task, which involved a classroom, top-level performance required accurate depiction of all the objects in the classroom, including precise alignment of rows and columns of individual desks. For the map-following task, which involved using a map of a school playground and following a route with four right-angled turns, top-level performance required making the correct turns at the precise points indicated by the distances and reference objects shown on the map.

Case and Okamoto found that performance on these spatial tasks was definable by a common underlying factor. Performance on each task was scoreable as being at one of the hypothesized four stages and mean performance on each task conformed to predicted age norms. In addition, in a case study, they found some evidence that training on the drawing task led to improvement in the other spatial tasks, not evident in two control subjects, although number knowledge was unaffected.

These linkages of map-drawing and map-following skills to other skills involving the use of frames of reference are intriguing. Much more empirical work is needed to confirm that the correlations and training effects are specific to the spatial tasks, with other tasks such as number understanding constituting a separable domain. If the findings are confirmed, a further goal would be to explore the relation of dimensional development, as operationally defined by the Case and Okamoto task set, to what may (or may not) be distinct lines of spatial development, such as establishing alignment. At present, there is little linkage between existing work on map and model understanding in general and the specific issues of importance to Case's central conceptual structure theory. Exploring linkages would be important to both traditions. Case and Okamoto need to be sure they have not simply selected spatial tasks peculiarly amenable to a dimensional analysis (otherwise, they can make no claim to domain generality, let alone to an overall theory of cognitive development). Investigators interested in children's understanding of maps may find that the dimensional analysis provides a helpful framework for integrating research and thinking.

Summary of Developmental Description

Children begin to be able to use simple maps and models in simple situations by the age of 3 years. That is, they can use maps presented to them in alignment with the real world to learn about the associatively coded placement of objects denoted by somewhat iconic, and even arbitrary, visual elements. In addition they have some ability to learn about distances in simple situations and to coordinate oblique and eye-level views of a space. However, over the next five or even ten years, map- and model-reading skills greatly improve. Interpretation of scale becomes increasingly more precise as measurement and calibration skills grow and children maintain scaling over multiple relations. Alignment between map and world becomes easier (but never simple) to establish as strategies for alignment are found and the efficiency of mental rotation improves. Planning and comparison skills emerge that allow for the use of maps in multistep navigation tasks. In addition advances in dimensional analysis and coordination have been suggested as possibly underlying development of mapping skills.

Map Competence and Developmental Theory

The development of map understanding has been analyzed from the perspectives of nativist, Piagetian, and Vygotskyan theory. The data we have reviewed demonstrate some strengths and some inadequacies in all three approaches. Here, we discuss the problems, as well as evaluating what each theoretical approach has to say about the role of input in this domain of cognitive development.

Piagetian Theory

Liben and Downs (1986) have taken a Piagetian position in which map understanding is said to be a late and hard-won achievement and, more important, an achievement not possible until underlying changes in spatial representation have taken place (i.e., the construction of projective and Euclidean systems of reference, and the acquisition of proportionality). In this chapter we have argued that the developments in understanding of alignment and scale occurring over the early elementary school years do not require a Piagetian explanation and in fact are better explained in other terms. Dealing with scale requires the consistent use of a spatial analogy over multiple terms. Dealing with misaligned maps requires dealing with conflict in frames of reference rather than construction of a whole new system of spatial representation. These abilities are central to map competence but do not require one to posit underlying representational change.

The Piagetian treatment of the role of input in cognitive development is generally as the soil that supports growth to a mature state. Children are thought of as seeking out the relevant information they need for cognitive advance, perhaps as a plant might take up more of a trace mineral from the soil were it to need that mineral for growth at a particular time. In thinking about the domain of map understanding, however, there is a problem. Children and adults are exposed to maps and models, even if spottily and unsystematically. If they could "take up" information as they needed it in this domain, it is hard to see why they would not achieve fuller competence, at least by adulthood.

Nativism

Landau (1986) and Blaut and Stea (Blaut 1997; Blaut and Stea 1971, 1974) argued that there is an innate basis for map understanding. However, the results they report fall far short of a demonstration of innate competence, in either of two senses of "innate," namely present at birth or acquired early and easily.

First, there are no data to support the claim that basic elements of map competence might be present at birth, and there is reason to doubt that they are. The findings cited by Landau and by Blaut and Stea to support their position simply indicate, in concurrence with other studies, that children as old as 4 years of age can establish element-to-element correspondence between maps and the world. Additionally their results suggest that children can establish angular correspondence between directions on a map and directions in the world when alignment is established for them by the experimenter. However, at least for element-to-element correspondence, younger children do not show these abilities (e.g., DeLoache 1987).

Second, while a more sensible construal of a nativist position on map competence may be that map skills show early and easy acquisition rather than presence from the start, the data reviewed in this chapter show that many aspects of map use (e.g., good control of alignment and scaling) are acquired slowly and gradually. They seem in fact to be imperfectly acquired even by adulthood. Five-year-olds simply do not show facility with maps and models in any way comparable to their facility with language (another ability said, with some justice, to be acquired early and easily).

Nativists generally treat environmental input as a simple "trigger" to cognitive advance. In analyzing the case of the development of competence with maps and models, if nativists were to accept that development in some aspects of competence is protracted and the adult state is far from ideal, they would have to argue that the environment fails to provide adequate triggers to development. But experience with maps, while often spotty and unsystematic, is not completely lacking, and hence such an argument is unsatisfying. Another approach would be to argue that the aspects of map use that develop slowly are not fundamental to the domain but rather epiphenomenal: that attentional deficits or planning deficits or problems with mental rotation explain children's development of mapping abilities and adults' imperfect control of them. These abilities, the argument might continue, are not part of the core definition of map competence. This move is a more plausible one. But the distinction between core abilities and peripheral ones would have to be made much more rigorously than anyone has attempted so far, and the core abilities shown to be present earlier and/or without the need of environmental input other than the ubiquitous sort provided by living in the physical world.

Vygotskyan Theory

Some studies, inspired by a Vygotskyan approach to cognitive development, have shown that various aspects of map use are more mature in real-world contexts familiar to people than in unfamiliar or decontextualized situations (Gauvain, 1993). While such findings could be interpreted in many ways, they are certainly compatible with the view that maps and models are cultural artifacts, which need to be explained and supported through apprenticeship in their use. The same conceptualization has inspired studies of the transmission of navigational systems in sea-faring peoples (Gladwin 1970), and theorizing about the effects of map use on spatial cognition (Uttal 1999).

The Vygotskyan approach views environmental input as scaffolding for cognitive development. Scaffolding has several desirable connotations as a metaphor for input in the context of map learning. One thinks of the usefulness of scaffolding as more extended in time than is implied in the nativist trigger metaphor for input, and one thinks of its use as more deliberate and more intended by the participants than is implied in the Piagetian soil metaphor. Maps and models may be like writing systems or formal mathematics: building on innately unfolding skills, as writing systems build on language or mathematics builds on number understanding. Mapping is unlikely to be spontaneously re-invented by individual children, it is rather transmitted through teaching interactions.

The ages of attainment of component skills to map competence described in this chapter can be argued, within this perspective, to be lower bounds on what might be achieved given better instruction. Children exposed to maps more extensively and systematically might acquire skills at younger ages than is seen in our current culture. The effects of such instruction are one of the unexplored areas in the intersection of cognition and instruction research, and are of some practical importance. A crucial question in geographic education concerns when the cognitive abilities of children allow for them to benefit from map-based instruction. Theoretical differences among investigators lead to answering this question differently (e.g., Blades and Spencer 1994; Blaut 1997; Downs, Liben, and Daggs 1988; Natoli et al. 1984). From our perspective, there are likely to be some limits in map instruction, but not ones based on qualitatively distinct modes of spatial coding. Rather, the building of strategies to deal with spatial problems requires cultural support: sequenced and guided instruction in the various components of map use. Further knowledge on the limits and possibilities of map instruction requires more systematic manipulation of input regarding mapping skills than has occurred to date, and observation of whether children's understanding can grow from this exposure. For instance, one might explore how teaching of measurement techniques relates to improvement in scaling on maps or, even more directly, how teaching of the principles of scaling improves scaling. Precise scaling requires mastery of multiplication and division, so one might examine also the reciprocal relations between development of these mathematical skills and refinement of control of scale in maps. The fact that spatial practice and input leads to spatial growth (Baenninger and Newcombe 1989, 1995; Huttenlocher et al. 1998) makes one optimistic that education can improve map skills, even for people who think of themselves as having difficulty with map tasks.

While we have advocated incorporating Vygotskyan ideas into future research on maps, that does not imply that we are thoroughgoing Vygotskyans with respect to all developmental issues. Our arguments here are specific to the use of maps and models. Maps and models seem to be artifacts or tools that, like writing systems and calculus, have been invented by particular human beings with special skills and insights. They build on innate human abilities but do not themselves constitute such abilities. Hence efficient learning of such symbolic and reasoning systems requires carefully sequenced cultural support.

Chapter 7
Space and Language

People routinely use language to describe space and to comprehend the descriptions of others. Every day we ask and answer questions about location and direction. For example, we say, "Have you seen the coffee filters?" and learn that they are "at the top of the pantry on the right," or inquire "How do I get to your office?" and find that we should, "Go north on Lancaster Avenue, turn right at the supermarket, and park at the red-brick building on your left." Responses to questions about location save us a good deal of time that might otherwise be spent in unguided search. Descriptions of location also serve to establish a sense of place in novels, and to provide a framework for understanding in nonfiction. Travel guides, for instance, supplement their maps with verbal descriptions of sites to see and places to visit (e.g., "straight ahead as you enter the tomb, you will see. . . . ").

While comprehension and production of spatial language is a common occurrence in everyday life, spatial language does present some challenges. In describing space, people must wrestle with at least three issues. First, distance and direction knowledge is encoded in memory in a fine-grained fashion with a fair degree of metric precision, as we have seen in previous chapters. Single words, however, generally allow people to capture spatial quantity only categorically (and hence inexactly); more exact communication requires either the use of quantitative terms which may be overly technical for everyday speech or the use of long and complicated phrases. Second, there are various plausible frames of reference used in spatial coding, and in specific situations there are usually multiple referents with respect to which location may be coded. Mental representation of space allows codings using various referents to coexist, providing both redundant and complementary information. One may remember, for example, that the soccer ball was placed on the grass on the left side of the red van, and also remember that it was about five feet down the road from the slide and between the water fountain and the bench. In speaking, however, people usually choose among referents and frames of reference, using only one system rather than listing all the spatial relations they know. Therefore they must make sure their listener knows what they intend as the frame of reference. Third, in talking about space, people must encode spatial relations in the linear one-at-a-time fashion required by language. In order to do so, they must not only select a subset of spatial relations which they think are most crucial for their listeners to know, they must also construct an order in which to present these selected spatial relations.

To comprehend spatial description, people must perform the inverse operations to those performed by speakers. Listeners must supply missing metric values for distances or directions only described in categorical terms, they must decide what frame of reference the speaker (or writer) has used, and they must integrate sequentially presented spatial relations, to form a coherent view of the spatial world. If communicators do a good job of supplying metric information where it is crucial, of selecting and specifying referents, and of ordering the spatial relations they mention in a logical way, comprehenders have an easier time acquiring spatial information. Still, even in the best of circumstances, the task of comprehension is probably never completely trivial. It takes effort to use sequential presentations of selected categorical spatial relations to construct a representation of an integrated and metrically coherent spatial world.

Speakers and listeners call on more than their spatial knowledge in order to make the choices and decisions required to communicate verbally about space. Success at the task also requires a good understanding of social interaction and cultural convention. Speakers must know how their linguistic and cultural community usually talks about space. They must anticipate possible misunderstandings and try to monitor for lack of comprehension in listeners where this is possible (e.g., in face-to-face situations, they may carefully observe facial expression). Like speakers, listeners must know the conventions, and must monitor their own developing understanding for ambiguities and possible misinterpretations so that they can ask questions if that is possible.

When we study the development of linguistic communication about space, we are thus really studying a multidimensional kind of development, consisting of several interacting story lines. The development of effective spatial communication is a function of spatial knowledge, of linguistic skill, and of social understanding and self-monitoring. Development along each of these lines is likely strongly supported by species-typical endowments in spatial processing, language, and social interaction. However, children still need to interrelate skills in the three domains and form them into a smoothly functioning package allowing for effective spatial communication. In this chapter we discuss what is known about comprehension and production of spatial language, beginning with studies of adult competence and then examining studies of children's acquisition of the relevant abilities.

In pursuing this agenda, we touch at several points on the question of the relation of language and thought. To what extent does the nature of what can be linguistically communicated about space tell us something about the nature of spatial knowledge? Does the particular means one's language uses to encode space affect the way one encodes spatial location in perception and memory? Does it affect the acquisition of spatial concepts? While the very strong position that the nature of language absolutely determines the nature of thought is widely disbelieved today, weaker versions of Edward Sapir's and Benjamin Whorf's position are still quite viable (Hunt and Agnoli 1991). Recently there have been several theoretical proposals regarding aspects of the language and thought question in the spatial domain. It has been argued that the nature of language can be used to draw conclusions about the nature of spatial coding (Landau and Jackendoff 1993), that people encode space differently, depending on the nature of the spatial terms in their language (Levinson 1996; Pederson 1995), and that children acquire spatial concepts at rates and in sequences depending on the spatial terms they hear (Bowerman 1996; Choi and Bowerman 1991; Gopnik and Meltzoff 1986).

Adult Talk about Space

In this section we deal with the three issues that create challenges for expressing spatial representations as linguistic descriptions: capturing metric coding in categorical language, choosing and specifying frames of reference, and ordering descriptions of spatial relations in a sensible linear sequence. In addition we take up two specific issues in the relation of spatial language to spatial thought. We argue against the idea that the categorical nature of spatial language indicates that spatial coding lacks metric precision (Landau and Jackendoff 1993), and we indicate some reservations about the claim that the nature of one's language affects what spatial relations in the world can be coded (Levinson 1996).

Metric Coding and Categorical Language

Single words, with the exception of proper names such as "Niagara Falls" or "Abraham Lincoln," refer to a grouping of similar objects, actions, or situations (e.g., cups, jumps, parties). Language typically abstracts away from specificity; using single words requires grouping phenomenal experience into frequently experienced clusters. So words, by their very nature, are not perfectly suited to the communication of fine-grained spatial information, which is intrinsically continuous. We can, however, easily use single words to capture the general area in which something is located (i.e., to specify a spatial category). For example, we say that the office is on your "right." In other words, although people represent spatial location both at a fine-grained level and at a categorical level (Huttenlocher, Hedges, and Duncan 1991), one-word (monomorphemic) references to space do not refer to fine-grained information but rather refer to membership in spatial categories (i.e., they place a location within a general area rather than precisely).

If speakers believe that single words do not code spatial location with sufficient precision, they have the option of supplementing this information with longer phrases or technical language in order to specify location exactly. One can say that "the coffee filters are in the pantry" or one can say that "the coffee filters are in the pantry, on the top shelf at the right, three inches from the teabags." There are at least two points worth making about such supplementation. First, the option of adding information to single-word references is not specific to space. When, for instance, a noun refers to a class of objects, but speakers think more specific reference is required, they can use adjectives or longer phrases to communicate more exactly what particular object they have in mind (e.g., instead of referring to "the cup," they may say "the china cup with the tulips on it that your Aunt May gave you"). Second, while supplementation results in exactness of reference, it is not always desirable. Referring to something at a categorical level using single words has advantages: it's simple and easy and generally sufficient for most common communicative needs. For instance, knowing that the coffee filters are at the top right of the pantry is adequate information to find them; both speakers and listeners would be unduly burdened by routinely expressing the exact location of the filters in metric coordinates.

Several questions arise about the nature of the spatial categorization inherent in the linguistic encoding of continuous spatial dimensions. What kind of categories are used? Do spatial categories have a graded structure, with prototypic instances and fuzzy boundaries, as do many natural categories? If so, how are areas of space grouped into categories? That is, what are the category prototypes, and where are continuous dimensions cut to create category boundaries (albeit possibly fuzzy ones)? What continuous dimensions are considered, and are some more psychologically salient than others? Do languages vary in how they categorize spatial relations, and what cognitive consequences does such variation have?

Research on the structure of spatial-linguistic category terms has generally focussed on small semantic fields in which words are poten-

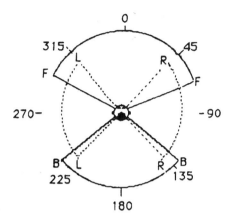

Figure 7.1
Mean boundary locations of eight conceptual regions around oneself, taken from experiment 1 of Franklin, Henkel, and Zangas (1995).

tially mutually exclusive. Such regions include "front," "back," "left," and "right" for the region surrounding the self (Franklin, Henkel and Zangas 1995), "above," "below," "left," and "right" for the relations of two objects shown on a plane surface (Crawford, Regier, and Hutten-locher, in press; Hayward and Tarr 1995; Maki, Maki, and Marsh 1977; Munnich, Landau, and Dosher 1997), and "north," "south," "east," and "west" for directions defined with respect to the earth's surface (Loftus 1978; Maki, Maki, and Marsh 1977). These studies have shown that people do not construe sets of spatial terms as mutually exclusive and precisely defined, nor do they treat locations within a category as equivalent. Rather, people seem to organize spatial terms around prototype locations, with graded degrees of membership.

Several kinds of behavioral data, as well as linguistic data (see Talmy 1983), lead to this view of spatial categories. First, people often con-catenate spatial terms in free descriptions, saying such things as "it's in front, but a little to the left" (Franklin et al. 1995; Hayward and Tarr 1995). Such concatenation is inconsistent with mutual exclusivity. Sec-ond, categories also overlap in comprehension (i.e., even when people are not asked to provide descriptions). In one experiment Franklin et al. asked people (seated in a homogeneous circular space) to point to the location that would be "as far to the right as possible so that you would still consider yourself as pointing to the front" (and all other possible permutations of the four terms). The results are shown in figure 7.1. There was considerable overlap among regions. "Front" was found to be a large region, extending considerably beyond the 90 degrees occupied by "right" and "left," and "back" was also larger than

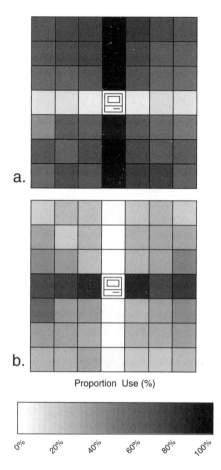

Proportion Use (%)

0%   20%   40%   60%   80%   100%

Figure 7.2
Proportion use of spatial prepositions in a production task for (a) vertically oriented spatial prepositions and (b) horizontally oriented prepositions. From Hayward and Tarr (1995).

90 degrees. Third, locations at category prototypes are rated as most appropriately described by a specific term, with locations further from the prototype rated as less and less appropriate (see figure 7.2, showing data from Hayward and Tarr). Fourth, category prototypes are used to understand less prototypic locations; Loftus (1978) found that people understand compass directions by computing the nearest cardinal direction and thinking in terms of deviations from it.

Spatial-linguistic categories are, then, like categories in other domains. They have prototypes, graded membership, fuzzy boundaries, and some degree of overlap rather than exhibiting strict mutual exclusivity. This general conclusion invites further investigation of spatial-linguistic categories aimed at understanding the determinants of where category prototypes and boundaries are located. A theoretically important goal of such study is to determine whether the structure of spatial-linguistic categories is congruent with, or different from, the structure of spatial-representational categories (as inferred from nonlinguistic behavioral tasks).

Hayward and Tarr (1995) have argued that the two types of categories are congruent. When describing two objects in relation to each other, Hayward and Tarr found that speakers use "above" to mean a prototypical location along a vertical dimension bisecting the referent and that they are also more accurate in reproducing or discriminating location along this vertical bisector (see also Munnich et al. 1997).

Hayward and Tarr's reproduction findings are, however, in apparent conflict with Huttenlocher, Hedges and Duncan's (1991) finding that when coding location of a dot within a circle, people use the vertically bisecting line, together with the horizontal bisector, to define quadrants as spatial categories, within which prototypes are centrally located (i.e., at roughly 45, 135, 225, and 315 degrees). Crawford et al. (in press) have recently shown that Hayward and Tarr's finding of greater accuracy in reproduction of locations on a bisector should not be taken to indicate the location of a nonverbal prototype. Although variability of responses is reduced for locations on the vertical axis, the direction of bias in reproduction is away from the vertical bisector, not toward it. So prototypes are located in the same areas found by Huttenlocher et al. (1991). This means that linguistic and nonlinguistic prototypes are not necessarily congruent. Along similar lines, Munnich et al. (1997) have argued that language need not constrain nonlinguistic spatial representation. They found that speakers of Japanese or Korean behaved differently than speakers of English in some spatial description tasks, while showing the same kind of memory for spatial location.

Another line of research on sets of spatial terms has compared the various dimensions in a spatial field to each other. Some dimensions are apparently more salient and simpler to process than others. Specifically, spatial terms labeling vertical dimensions on two-dimensional visual arrays ("above" and "below") are easier to verify than terms labelling horizontal dimensions ("left" and "right") (Maki et al. 1977). Similarly, verifying "north" and "south" relations is easier than verifying "east" and "west" relations (Loftus 1978). In the three-dimensional world several studies have shown that upright observers find their head–foot axis most accessible, likely because that axis is both asymmetric and gravity defined. The front–back axis is intermediate in difficulty; it is asymmetric, but not gravity defined. The left–right axis is the most difficult. The situation changes somewhat when observers are reclining. In this case people find the front–back dimension easiest to process, followed by the head–foot dimension. Left–right remains the most difficult of the three dimensions (Bryant and Tversky 1992; Franklin and Tversky 1990).

The spatial categories we have discussed so far are all members of restricted semantic fields in which location is defined with respect to two- or three-dimensional systems. But of course there are many spatial terms that are not so simply defined but whose meaning also involves considerations of support, containment, type of fit or attachment, and so on. The importance of these dimensions is especially striking when we consider cross-linguistic variation in which dimensions are considered to define spatial categories (Bowerman 1989, 1996; Landau 1996). For instance, in English, the word "on" is used to refer to horizontal support (e.g., "the cup is on the table"), to vertical support (e.g., "the picture is on the wall"), and to encirclement (e.g., "the ring is on the finger"). Thus certain potential distinctions are ignored. However, three different prepositions are used in German for these three relations. Such linguistic variations are often taken as the departure point for arguing that language exerts an influence on thought. An alternative hypothesis is that spatial relations are conceptualized similarly by speakers of different languages, with languages imposing different kinds of categorical organization on these underlying continuous and metric representations. This controversy is discussed shortly in the context of exploring acquisition of these spatial terms.

In this section we have developed the view that spatial coding occurs on both fine-grained and categorical levels, with linguistic terms mapping onto categorical codings, although not always in a direct way. This view is quite different from Landau and Jackendoff's (1993) claim that the categorical nature of spatial language suggests that spatial information is encoded only categorically. Landau and Jackendoff point out that there exist a relatively few terms (for English, mainly prepositions) encoding spatial relations between objects, especially as contrasted with the many words existing to code object shape (essentially, all concrete nouns). They suggest that this contrast tells us something about the nature of the "what" and the "where" systems (see Unger-leider and Mishkin 1982).

The Landau-Jackendoff hypothesis seems odd to us, for two reasons (and Landau, at least, seems recently to have backed off from it; see Munnich et al. 1997). First, accepting the hypothesis requires one to disregard the fact that empirical studies of people's ability to remember spatial location show an impressive degree of fine-grained accuracy.

Such accuracy is evident both in studies of spatial memory in adults (e.g., Huttenlocher, Hedges, and Duncan 1991) and in children as young as 16 months (Huttenlocher, Newcombe, and Sandberg 1994). Accuracy is also evident in considering the precision of the motor system in locating objects or in skilled athletic performance (Hoffman 1993; Tversky and Clark 1993). Given these facts, coarseness in the spatial linguistic system does not seem likely to reflect a fundamental limitation on the nature of spatial representation. Rather, as we have already argued, such relative coarseness is a common feature of most kinds of language and likely reflects the need for economy in communicative situations.

While the empirical findings on fine-grained spatial memory seem to us sufficient reason for questioning Landau and Jackendoff's hypothesis, there is a second problem with their idea as well. Their argument depends on comparing the categorical coding of spatial information in language to the categorical coding of information regarding objects in language. Spatial information is communicated in English using prepositions, a closed class with around a hundred or so members. Object information is communicated using nouns, an open class with a very large number of members. Landau and Jackendoff's comparison of the sizes of the two classes, which they use to support the idea that spatial information is only coarsely coded, is based on the proposition that nouns encode spatial information about objects. Basically they are saying that it is curious that we have only a hundred or so terms to refer to location but thousands of terms to refer to shape. But do nouns really encode fine-grained spatial information? Nouns do refer to classes of objects defined in part by their form, extension, and shape, but as Slobin (1993) puts it, nouns refer to things, not shapes. Using a count noun does not encode the shape of the object to which the noun refers in any precise way, and such a word encodes many other things as well (e.g., the object's function). There are many words referring to things having roughly the same shape (e.g., small mammalian quadrupeds), but this is not to say that we recognize many fine gradations in shape; the contrasts between the words may also depend on the animals' other characteristics, such as domesticity or diet.

These two arguments show that it is unwarranted to infer limited or purely categorical conceptual coding of spatial relations from the linguistic fact that spatial relations are encoded by a closed class of words with categorical scope. Given direct evidence that location is encoded in a fine-grained as well as a categorical way, categorical linguistic encoding of spatial location is best seen as part of the nature of language, justified by its relative simplicity and its adequacy for many communicative situations. People always have the option of supplementing monomorphemic specification of spatial relations with lengthier phrases or the use of conventional measurement units when such fine-grained specification is required. It is hard to see how they could do this if they had not coded more information than they routinely express using single words.

Frames of Reference

The second problem people must solve when they talk about space is to avoid ambiguity in using landmarks to describe the location of a target object. For instance, shown a chair viewed from the side, and asked to put the ball "in front of" the chair, listeners may put the ball at the chair's front, if they take the chair as the referent and interpret "front" as meaning the functional side of the chair. Or, they may put the ball between themselves and the chair (i.e., to the chair's side), if they take themselves as the referent and interpret "front" as meaning in the direction that they're facing. Or, if they are speakers of Hausa, they may put the ball on the side of the chair most distant from themselves, conceptualizing themselves and the chair as aligned along a dimension on which the ball should be the leader if it is in "front."

The fact that there may be more than one meaning for a seemingly simple spatial description would seem to threaten speakers and listeners with frequent communicative dilemmas. Recent research has shown, however, that several factors help speakers to avoid such problems and to communicate with listeners. First, it must be remembered that different frames of reference are not always in conflict. When frames of reference agree, their congruence facilitates verification of spatial statements (Carlson-Radvansky 1995). Second, speakers and listeners (at least adult ones) are quite flexible in their use of different frames of reference and realize that ambiguity is a problem in these communicative situations. Conversational partners negotiate their choices of frames of reference as their interaction progresses. When they switch speaking and listening roles, they generally maintain the choices of reference frames already established. They also consider it more polite to use a nonegocentric perspective than to speak egocentrically, although egocentric frames of reference are used more frequently with real conversational partners (who may license use of such frames and indicate they understand them) than when descriptions are being generated for hypothetical listeners (Schober 1993, 1995). Third, there are certain conventions governing people's choices of frames of reference. For instance, when there are functional relations between two objects (e.g., a person reaching toward a mailbox), people prefer to use intrinsic descriptions (i.e., "the person is in front of the mailbox"), but when the objects are not in a functional relationship (e.g., a person with her back to a mailbox), viewer- or environment-centered descriptions are preferred (e.g., "the person is to the left") (Carlson-Radvansky and Radvansky 1996). Fourth, once a certain reference frame has been activated, other possible reference frames seem to be inhibited, thus easing the problem of interpretation (Carlson-Radvansky and Jiang 1998).

There has been considerable discussion in the literature concerning the merits of various competing taxonomies for the analysis of linguistic frames of reference (see chapters in Bloom et al. 1996 for a full review). One proposal is that spatial descriptions can be considered as viewer centered, as object centered (defined with respect to the intrinsically defined sides of a fronted object), or as environment centered (Fillmore 1975; Miller and Johnson-Laird 1976). Within object-centered descriptions Miller and Johnson-Laird also suggested that describing location with respect to an object might be easier than describing location with respect to an addressee, an idea that has recently received empirical support (Schober and Bloom 1995).

An alternative taxonomy is presented by Levinson (1996), who suggests that "frames of reference" in linguistic description of space can be most sensibly grouped into a different set of three categories: intrinsic, relative, and absolute. Using an intrinsic frame of reference, a speaker indicates the location of a target object with respect to a single referent with inherent features, often a front, back, and sides. This referent may be the self, another person, or an object or building with an intrinsic structure (e.g., a chair). Using a relative frame of reference, a speaker indicates the location of a target object with respect to two referents. In some cases, one of these referents is either implicitly or explicitly the viewer, as when one says "the ball is to the left of the tree," a phrase that has meaning only if the listener knows the position of both the speaker and the tree. In other cases, both referents may be external objects, as when one says "Bill kicked the ball to the left of the goal," locating the ball with respect to Bill and the goal. Using an absolute frame of reference, a speaker indicates the location of a target object using a system of fixed directions referring essentially to points of the compass, sometimes as embodied in geographic features tied to points of the compass such as slope of the ground or prevailing wind direction. (See figure 7.3.)

The viewer/object/environment taxonomy differs most notably from the intrinsic/relative/absolute taxonomy in whether a viewer versus other classification is seen as fundamental or whether the number of differentiated referents is taken to be the more important distinction. It remains to be determined whether one approach gives investigators more purchase than the other over describing and

INTRINSIC

"He's in front of the house."

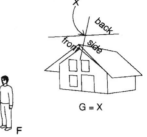

G = X

F

RELATIVE

"He's to the left of the house."

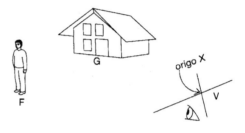

F

origo X

V

ABSOLUTE

"He's north of the house."

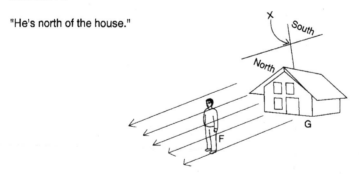

Figure 7.3
Canonical examples of the three linguistic frames of reference proposed by Levinson
(1996).

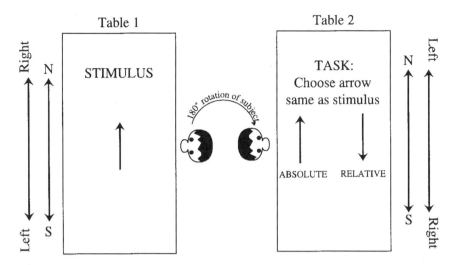

Figure 7.4
Underlying design of the experiments reported by Levinson (1996).

predicting the choices speakers make. However, we wonder whether these issues will turn out to be crucial. People can use the self, other people, or fronted objects as referents, and they can use these referents alone or in various combinations. In addition they can describe spatial location in viewpoint-independent terms in some cases (e.g., if objects are "between" two landmarks or can be located "in the middle of" a definable area), and they can also use compass directions or their equivalents. In short, there may be more possibilities for linguistic description of space than three, and neither the viewer-other distinction nor the one- versus two-referent distinction seems particularly likely to be psychologically privileged over the other.

Cross-linguistic research has recently identified an intriguing phenomenon concerning the relation of people's linguistic frames of reference to their spatial coding. Speakers of European languages generally find that use of an absolute frame of reference (i.e., compass directions or their equivalents) has a technical flavor and seems difficult to use habitually. For instance, it seems odd to an English speaker to say that "the coffee cup is to the north of the sugar." However, certain languages in the world emphasize use of such absolute frames of reference and even require them. Levinson (1996) has looked at processing of simple situations such as that shown in figure 7.4 by English or Dutch speakers and by speakers of languages requiring absolute linguistic reference, such as Tzeltal, a Mayan language spoken in the Yucatan, and Guugu

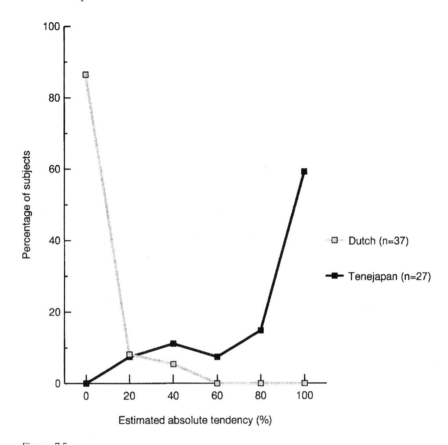

Figure 7.5
Responses of different language communities on the animals recall task. From Levinson (1996).

Yimithirr, an indigenous Australian language. After reversing orientation, when asked to reproduce the stimuli, English and Dutch speakers very consistently reproduce the relative locations of the pair, while speakers of Tzeltal or Guugu Yimithirr retain their absolute location, reversing their relative location (see data in figure 7.5). Similar findings have been obtained by Pederson (1995), who compared performance by speakers of two closely related varieties of Tamil, thus controlling for the wide variety of nonlinguistic features that differ between European and Mayan or Australian informants.

One interpretation of these findings is that they provide evidence that language indeed influences thought. Or, to put it less baldly, it may be argued "that the underlying representation systems that drive all these capacities and modalities have adopted the same frame of refer-

Box 7.1

In the following passage, taken from a popular children's book, three people are looking down from a castle at a message "UNDER ME," which is deeply inscribed on the stones below. The speaker is trying to describe to the others how they had "seen" the letters on passing over the ground the day before, without realizing what they were doing.

"Why, you chump!" said Scrubb. "We did see it. We got into the lettering. Don't you see? We got into the letter E in ME. That was your sunk lane. We walked along the bottom stroke of the E, due north—turned to our right along the upright—came to another turn to the right—that's the middle stroke—and then went on to the top left-hand corner, or (if you like) the northeastern corner of the letter, and came back. Like the bally idiots that we are."

From C. S. Lewis (1953/1981), *The Silver Chair*, p. 123. Harpercollins Juvenile Books.

ence" (Levinson 1996), presumably as a result of the influence of language on thought over developmental time. There is, however, an alternative way to think about these demonstrations. Simply, people in these experiments are faced with an ambiguous situation. There are two possible ways to interpret the experimenter's instructions, and participants must essentially guess which meaning the experimenter intends. Given this ambiguity, people interpret the situation in the way that is more plausible in their linguistic conventions. On this view, informants might, almost as easily, pick the alternative interpretation, and they certainly are not incapable of understanding an alternative possibility or of conceptualizing space in the alternative way.

Consider a simple thought experiment. Suppose that one said to English speakers that the task is to retain the positions of the objects relative to the points of the compass. Would the relative rarity of such usage in everyday English prevent people from following the instruction (assuming they knew which direction north was)? This is an empirical issue, but it seems unlikely that they would fail. Certainly, in a task involving giving directions from a city map, even speakers who did not naturally use points of the compass found it quite easy to do so when asked to use cardinal directions (Ward, Newcombe, and Overton 1987). It seems even more likely that an analogous instruction (i.e., to use relative terms) would be successfully followed by speakers of Tzeltal or Guugu Yimithirr, given the fact that the data shown in figure 7.5 indicate less extreme linguistically influenced bias in their interpretation of the ambiguous command to start with.

In short, people do not seem to need to commit themselves to one linguistic frame of reference only. In fact they fairly frequently mix frames of reference in a single passage, perhaps to maximize the chances of listener comprehension (Taylor and Tversky 1990 and as illustrated in box 7.1). The fact that people have many alternative strategies available to talk about space suggests that they are not compelled by the characteristics of their language to abandon methods of spatial description not frequently used by that language. Facts about language influence strategies for linguistic communication of spatial information, not spatial information itself.

Sequentiality and Simultaneity
The third problem people face when they try to talk about space is that they must order spatial relations sequentially, since they can only say one thing at a time. How do speakers choose an order in which to mention the various spatial relations they know? One strategy is to "give a tour." That is, one can begin at a logical entry point, such as the front door of a house or the parking lot at the head of a system of hiking trails, and then tell the listener about each of the landmarks that would be seen along a specified route, described in a viewer-centered way (e.g., "on your left, you will see . . . ") This strategy is the one used by most people in Linde and Labov's (1975) study of descriptions of apartments.

But tours are not the only way to structure spatial descriptions. People may describe location in terms of a bird's-eye view of an area, giving a "survey" of the space. For example, they might describe a town by saying that the post office is north of the drugstore, with the hardware store across from it. Survey descriptions are generally no less structured than route descriptions. That is, people do not simply describe elements in a random order from a bird's-eye view. Instead, they tend to structure their descriptions hierarchically. Sometimes they begin at the top of a hierarchy, starting by describing the general region or the most important landmarks and then filling in details. For instance, Shanon (1984) asked people to describe their dormitory rooms and found that they first described the large stable elements and then mentioned small movable objects. The opposite order can be used as well. People sometimes begin survey descriptions by mentioning the location of local landmarks (e.g., "the keys are on the table"), then proceeding to provide information on increasingly larger regions or landmarks (e.g., "the table that's in the hallway next to the dining room door").

An important factor determining whether ascending or descending the hierarchy is a preferable way of structuring survey descriptions seems to be whether one is giving directions or simply describing. People usually start directions with the more general unit and descend the hierarchy, whereas descriptions are more likely to begin with a local description and ascend the hierarchy (Plumert et al. 1995). Presumably this difference is based on pragmatics; it is more helpful to listeners who need to go somewhere to get them first to the right room, and then help them to focus on a particular location, because that is the order in which they will execute the actions. On the other hand, listeners who are in need of descriptions may frequently have an attentional focus, on a character or an object in a narrative, and they can benefit from a spatial description that builds up from that center of attention.

What determines whether speakers will give a survey description of a space at all, or will prefer to give a route description? Taylor and Tversky (1992a) found considerable diversity. Speakers were more likely to give "survey" descriptions for a town and "route" descriptions for a convention center, while they divided evenly on which strategy they used for describing an amusement park. A subsequent experiment investigated whether biasing factors in choosing a strategy for description were the overall size of the area, whether the space was enclosed or open, whether there was a single obvious path through the area, or whether or not the landmarks in the space were of roughly equal size (Taylor and Tversky 1996). The findings suggested that the first two factors (overall size, and enclosed versus open spaces) were unimportant, but the third and fourth factors were influential. Speakers were more likely to use route descriptions when the landmarks described were roughly equivalent in size and, perhaps not surprisingly, when there was an obvious path to structure the description.

So much for the speaker side of the equation. Given these variations in speaker strategy, it is interesting to ask whether listeners can use the ordered descriptions produced by speakers to construct coordinated spatial representations, and whether the ease and accuracy of doing so is affected by the strategy adopted. The answers to these questions seem to be "yes" and "no," respectively. First, listeners (or readers) can take separate statements regarding spatial relations and integrate them into a single representation of the state of affairs described. This phenomenon was in fact at the heart of many of the demonstrations of "semantic integration" in the early 1970s (Barclay 1973; Bransford, Barclay, and Franks 1972). Integration is difficult when the set of statements do not unambiguously fix the relative positions of the elements described, however, and in this case listeners are more likely to retain verbatim memory for the input sentences, perhaps to aid in disambiguation should further information become available (Mani and Johnson-Laird 1982). Second, experiments comparing how people who listen to or read route descriptions versus those who listen to or read survey descriptions answer route versus survey questions have repeatedly shown that there is no difference in performance (Taylor and Tversky 1992a, 1992b). In fact, people who study maps perform very similarly to people who acquire spatial information from verbal descriptions of either type, suggesting that the underlying spatial representation is similar despite variations in how the information was originally acquired (Taylor and Tversky 1992a, 1992b, 1996).

Research directed at determining the nature of the representations formed in the course of comprehending complex discourse provides further information relevant to the first conclusion. Several studies have demonstrated that readers find it easier to access objects located close to a protagonist than objects located farther away (Glenberg, Meyer, and Lindem 1987; Morrow, Bower, and Greenspan 1989; Morrow, Greenspan, and Bower 1987; Morrow et al. 1992). Such findings seem to support the argument already made from the semantic-integration literature, that listeners (or readers) construct coordinated spatial representations from linguistic descriptions. There has lately been some controversy about this matter, however. Rinck et al. (1997) showed that the reading time effect depended on categorical, not Euclidean, distance (i.e., on number of rooms traversed, not on the rooms' sizes). Langston, Kramer, and Glenberg (1998) also found that readers do not automatically make spatial inferences, and they concluded that the spatial representations formed from text may be more topological or functional than Euclidean.

The basic issue may concern the extent of effort required to construct metric spatial representations from complex discourse involving many locations. Rinck et al. (1997) found that people were able to use Euclidean distance to answer questions that could not be answered correctly without considering it. Thus the situation models used in comprehending text seem to include metric spatial information, although such information is used only when needed. In addition, even when situation models appear to be nonmetric, one wonders if this lack reflects the extent to which spatial working memory has been taxed during reading. Some people, especially those with less spatial working memory or those reading in conditions in which they have no reason to suppose they will need accurate spatial representations, may simply not be able to, or may not try, to form them. Evidence in support of this hypothesis has been provided by Friedman and Miyake (in press).

Summary
We started with the idea that using language to describe spatial relations poses three challenges: cutting continuous space so as to describe it categorically with single-word terms, selecting landmarks to use as

referents when a variety of possible frames of reference exist, and casting simultaneous relations into some sequential order. What we have found is that on the whole, adult speakers and listeners cope with these tasks remarkably well. They use spatial terms, as they use other categorical terms, to refer to areas with prototypic centers and somewhat unclear boundaries. They select frames of reference in a socially negotiated process with many pragmatic guidelines of which they are well aware. They order descriptions using defined organizational strategies, generally either by following a route through the environment or by giving a bird's-eye view in which each element is linked to the other in a systematic way. Which strategy is adopted matters relatively little, since listeners are quite adept at constructing integrated spatial representations from either kind of description.

Against this backdrop of adult competence, we can more easily pose the developmental questions of how these competencies are acquired. In particular, we want to know when and how children acquire spatial-linguistic categories, when and how they acquire the ability to negotiate frames of reference, and when and how they acquire organizational strategies to structure their verbal descriptions of space.

Acquisition of Spatial Language

Categorical Terms

As we have seen, describing spatial relations, at least in one-word references, requires speakers to group spatial locations categorically into regions with a central or prototypic member and other members of graded qualifications for category membership. Different languages have different spatial categories encoded in their spatial terminology. They may either require speakers to note certain aspects of a spatial situation, such as tightness of fit, or make it difficult to speak about that aspect in a simple fashion at all. How do children acquiring language learn what spatial categories are enshrined in the semantics and even syntax of their language?

One answer to this question is that spatial language depends on already acquired spatial understanding (e.g., Johnston 1985). That is, based on development in the first 12 to 18 months, children have acquired certain spatial notions and contrasts; the task of acquiring spatial language is then to map words onto these established conceptual distinctions. There are several arguments for this "semantic primitives" position, reviewed in detail by Bowerman (1996). First, the sequence in which children acquire spatial terms has seemed to many investigators to correspond to a sequence of spatial development.

Spatial terms are acquired beginning in the second year of life, in an order showing a fair degree of cross-language consistency. "In," "on," and "under" are among the first spatial terms to be acquired, along with directional terms along the vertical dimension ("up" and "down"). Terms encoding proximity, such as "beside" and "between" are acquired next, followed by terms involving specification of a frame of reference (e.g., "in front of" and "behind"). "Left" and "right" are notoriously late. Second, nonlinguistic knowledge generally seems to be discernible earlier than linguistic expression (e.g., Halpern, Corrigan, and Aviezer 1983). Third, early spatial words, most notably "up," are extended rapidly to a wide variety of contexts in which either vertical movement or vertical position need to be linguistically coded (Bloom 1973; Nelson 1974).

Recently, however, there has been vigorous discussion of whether nonlinguistic spatial understanding is actually sufficient to account for the acquisition of spatial terms. Bowerman (1996; see also Choi and Bowerman 1991) has argued that spatial terms are not mapped directly onto pre-existing spatial concepts but that, instead, the semantic organization of the input language affects the evolution of children's spatial concepts. In the spirit of Brown's (1958) description of the "original word game," Bowerman suggests that how adults use words, and across what contexts adult word use is invariant, provide important sources of information to children about how to organize their spatial world.

The impetus for this proposal comes from cross-linguistic study of how children express locational concepts. As we have already seen, different languages choose very different ways of dividing up the spatial world. For instance, English uses "in" to express containment either in a closely fit container (e.g., "the knife is in its sheath") or in a much larger container (e.g., "the strawberry is in the bowl"), whereas Korean distinguishes these kinds of containment, using different words for each. Furthermore the English word "in" applies both to situations using transitive verbs (e.g., "he put the knife in its sheath/the strawberry in the bowl") and to intransitive situations (e.g., "he lay in bed"), whereas Korean distinguishes the transitive and the intransitive situations, using yet more lexical items. Choi and Bowerman (1991) have shown that children acquiring Korean learn to express the contrasts marked in their language (e.g., between reference to tight-fitting containment and loose containment) at a pace roughly equivalent to that shown by English children in acquiring the contrasts of English. At the same time Korean children are slow to acquire terms that allow them to refer directly to concepts such as "up" and "down." These terms exist in Korean but are not central to the Korean language's system of spatial terminology. Choi et al. (1999) have shown that there are contrasts between Korean-acquiring and English-acquiring children in comprehension as well as production. Using a preferential looking technique, Choi et al. found that infants aged 18 to 23 months looked at visual events in patterns that indicated that the English-acquiring children were guided by the meaning of "in" in English, and the Korean-acquiring children were guided by the meaning of "kkita" in Korean.

These findings are clearly not fatal to a semantic primitives hypothesis, as pointed out by Mandler (1996). Advocates of the position can respond by suggesting that many spatial concepts are already available to the young language learner, whether or not the concepts will turn out to be important to linguistic distinctions. For instance, for the contrast between loose- and tight-fitting, which is an important one in Korean, Mandler suggests that children have sufficient experience with things fitting tightly (e.g., popping plastic beads into each other) and things fitting loosely (e.g., putting a plastic bead into a cup) that they might reasonably be expected to understand loose- and tight-fittingness. In Mandler's view, by the time children begin to acquire words, at the age of a year or so, they are equipped with a crucial set of nonlinguistic concepts; whatever they lack, they acquire in the next year or two. Their conceptual advances allow for subsequent linguistic acquisitions of the distinctions encoded by their particular input language. Such a process would be analogous to the way in which different languages build on, collapse, or simply fail to exploit the various phonemic contrasts that are innately available to the infant.

Bowerman (1996) acknowledges that Mandler's approach is not decisively refuted by currently available evidence, but argues that the semantic-primitives solution comes with a cost. The hypothesis postulates that all of the many contrasts used in spatial terms by the world's languages build on concepts available to the young language learner, and Bowerman suggests that there are too many such contrasts to feel optimistic that all of them are primitives. Perhaps a more powerful point is made by Choi and Bowerman (1991) who argue that a semantic-primitives hypothesis does not explain why Korean children, if they are equipped with the same nonlinguistic understanding of the vertical dimension as are English children, take much longer than English children to acquire words in Korean that directly express notions of upness and downness. Choi and Bowerman reject perceptual salience as an explanation, arguing that the relevant words are stressed in caregiver speech. The delay can be explained, they say, only if one assumes that language directs the development of understanding; these

Korean words are less central a part of the Korean language's system for coding location than are corresponding terms in English.

The difference between the semantic-primitives position and the linguistic-input position is fairly subtle, and may be difficult to adjudicate empirically. By the age of 5 months, infants show evidence of coding distance in continuous space (Newcombe, Huttenlocher, and Learmonth, in press), and by the end of the first year of life, children are able to use their metric location codings to find objects (Bushnell et al. 1995; Huttenlocher, Newcombe, and Sandberg 1994). So the problem of learning most spatial terms is basically the problem of deciding how to go about cutting these dimensions into categories. Some categories are present early, probably because they are perceptually natural. For instance, children as young as 16 months in the Huttenlocher et al. studies coded location categorically (as well as on a fine-grained level) as being "in the box," probably because the box is a bounded entity demarcated by a continuous uniformly colored and uniformly shaped edge. The category prototype was the middle of the box. Other categories take longer, sometimes much longer, to develop. For instance, children as old as 6 years do not spontaneously categorize the sandbox as consisting of two categories, a left half and a right half, an organization that is evident in 10-year-olds.

It seems plausible that obligatory or even frequent linguistic marking of a particular kind of categorical organization of fine-grained spatial coding might lead children to make categorical cuts they would not otherwise make, at least linguistically and perhaps in nonlinguistic spatial coding as well. For instance, imagine learning a language in which location was categorically described as either within 50 yards of a place where one would be safe from tigers (i.e., a relatively short sprinting distance) or outside that safety limit. If children consistently heard one term applied to people who were within a certain radius of a safe location, they might develop a conceptualization of their fine-grained spatial codings in which 50 yards radius was a salient and habitually made cut and in which "safe" locations were habitually used spatial referents. In this sense their language learning would be guided by the input (in the fashion of the "original word game"), and their nonlinguistic spatial coding system might even be affected by their language (a Whorfian finding, about which Choi and Bowerman actually declare themselves agnostic). Munnich et al. (1997) call this the accessibility hypothesis.

In this way of looking at things, prelinguistic spatial development lays the groundwork for acquisition of spatial language, in that children have acquired the ability to code location both along continuous dimensions and categorically before they need to tackle the task of language acquisition. However, the precise nature of their spatial categories undergoes considerable evolution over the next eight or so years. It would not be surprising if distributional facts about language use were in part responsible for directing this evolving categorical system.

Frames of Reference
We have seen that adult speakers choose frames of reference in a socially negotiated process that begins with the recognition that referents may be ambiguous and that incorporates certain rules about what kinds of referents are more appropriate when. People talking about space take advantage of conventions, and they also work to establish explicit agreement on how they will use spatial referents in a particular interaction. These processes allow for listeners and speakers to be sure that they understand one another.

Using linguistic frames of reference, either as a speaker or as a listener, requires spatial understanding, linguistic knowledge, and the ability to manage the social processes involved in communication. There are also sometimes general cognitive processing demands, for instance, on verbal or spatial working memory. Development of the ability to use linguistic frames of reference effectively could depend on development in any or all of these components. Research to date suggests that children's ability to produce and comprehend spatial referents depends on a blend of these components, although perhaps especially on changes in management of social interaction.

A good part of the spatial competence necessary to use frames of reference is present quite early in life. From a very early age, children code spatial location with respect to the self (one possible frame of reference used in talking about spatial location). By at least 21 months of age, children begin to code location using distal referents (another possible frame of reference used in communicating about spatial location) (Newcombe et al. 1998). In addition, from an early age children know the canonical fronts of objects and landmarks so that they have the conceptual basis to speak of an object as being, for instance, "in front of" or "behind" another object. Levine and Carey (1982) found that very young children, asked to form a "parade," could arrange objects such as shoes and chairs so that the objects all faced the same way.

There are, however, at least two kinds of spatial competence needed for some kinds of linguistic reference which may not be present in the first two years of life. First, the use of an absolute frame of reference in which spatial relations are expressed using terms corresponding to compass directions depends on the ability to maintain orientation to geocentric coordinates across a wide variety of environments. Because speakers of English are not required to encode such relations routinely, we know very little about the development of the ability to code location in this way. Studies of children acquiring languages such as Tzeltal or Guugu Yimithirr will be required to ascertain whether such systems are early or late acquisitions in cultures where children are exposed to relevant input. Second, children's ability to use the terms "left" and "right" correctly with respect to another person or fronted object may depend on strategies of mental rotation not easily or routinely implemented until 8 years of age or so. Although preschool children are able to identify left and right in visual displays using either viewer-centered or environmental frames of reference (Fisher 1990), identifying left and right from another point of view is clearly a more difficult task. Children show above-chance ability to identify what objects would be on the left or right sides of other observers by the age of 4 years (Newcombe and Huttenlocher 1992), but they do so at levels far from perfection, and need to be given instructions designed to maximize their understanding of the task. For instance, children had stickers put on one of their hands, and adults referred to the children's "sticker hand" and "hand without a sticker" rather than to "left" and "right." In addition an initial experience with a doll wearing a sticker on one hand and positioned in various locations around the table was used to give the children the idea of what kind of reference was intended by the terms "sticker hand" and "hand without a sticker." In more challenging situations, when the referents are more abstract (e.g., a triangular figure on a computer screen) and the instructions less careful, children as old as 6 years appear not to use mental rotation strategies to identify left and right (Roberts and Aman 1993).

A large part of what develops in the ability to use linguistic frames of reference may not involve spatial competence, however. For example, the children studied by Levine and Carey could correctly make a parade in which all objects faced frontward, well before they were able to place objects appropriately in response to directions to put them "in front of the chair" or "in back of the shoe." This fact suggests that, while the children knew how the objects conventionally faced, they either did not understand the use of the terms "in front" and "in back" to refer to certain features of the object or did not understand the terms as fixing spatial location in relation to such features. Similarly 3-year-olds can interpret "this" and "that" correctly in a situation with feedback (de Villiers and de Villiers 1974), but children as old as 7 years make many mistakes in interpreting the same words in a more formal testing situation without feedback (Webb and Abrahamson 1976). Children of 7 years, and even as old as 10 years, engaged in very little work on establishing common ground for their use of spatial referents in the same task in which adults studied by Schober (1993) did so easily and efficiently (Taylor and Klein 1994). Children of 3 or 4 years may fail to use their own position or the position of external landmarks, such as another speaker or a colored curtain, to describe location, even though we know that their own spatial codings utilize these referents (Craton et al. 1990; Plumert, Ewert, and Spear 1995). Such findings suggest that conceptualization of spatial location is established well before children master the cultural and linguistic conventions of how to refer to spatial referents.

Mastering the conventions is not a simple matter, since children hear many conflicting utterances. For instance, "in front of the shoe" may mean between the side of the shoe and the observer (a correct usage when the speaker intends a viewer-centered frame of reference) or at the tip of the shoe. Thus the developmental task in mastering deixis is not a matter of development of spatial representation but rather a matter of learning that more than one frame of reference is possible, and that frame of reference can be negotiated between speaker or listener (Schober 1993, 1995) and can also be established by factors such as whether or not the interobject relationship is a functional one (Carlson-Radvansky and Radvansky 1996).

Another important determinant of children's ability to comprehend even moderately complex spatial descriptions may be their verbal working memory. When given a few simple sentences describing the spatial relations of a few objects, children between the ages of 7 and 11 years show semantic integration; that is, they integrate the descriptions to form a cohesive spatial layout that allows for inference (Paris and Mahoney 1974). However, children of the same age, asked to listen to or read longer descriptions of more complicated spatial layouts, vary greatly in their ability to make spatial inferences about the layouts and to place objects on a map to represent the layouts (Allen and Ondracek 1997). A good predictor of individual differences at each age was a measure of verbal working memory, exactly what seems likely to have been stressed in dealing with passages like those used by Allen and Ondracek as opposed to those used by Paris and Mahoney. The role of working memory was also supported by a training experiment. When 6-year-old children were taught a strategy to limit working-memory demands, namely pointing to a piece of paper containing three direc-tional labels to "place" each object as it was mentioned in a narrative, subsequent ability to construct an actual map improved substantially (Ondracek and Allen 1997).

In summary, while spatial competence is well advanced by two years, when children first begin to confront spatial language in volume, there are some subsequent conceptual advances (i.e., in understanding absolute reference using compass points and in the mental rotation required to use "left" and "right") that may pace the development of communication using frames of reference. In addition changes in working memory may affect ability to produce and comprehend descriptions of complex scenes. In great part, however, the development of children's spatial communication seems to depend on their awareness of linguistic and social conventions and their ability to monitor listener comprehension and engage in communicative negotiation.

Children's Direction Giving

Adults organize their spatial descriptions either by following routes (the mental tour strategy) or by giving survey information which they hierarchically organize, proceeding either from general to specific or from specific to general, depending on the pragmatics of the situation. Children's ability to use such strategies will depend on their understanding that these approaches are helpful to listeners, as well as on whether their spatial representations are sufficient to allow them to access information in either a route-following order or a hierarchically organized order. While some theorists, beginning with Siegel and White (1975), have postulated that children's spatial representations are more likely to be route representations than to be survey representations, we have seen that there is evidence to question this hypothesis. Spatial representations may be hierarchically organized beginning by at least 16 months (Huttenlocher et al. 1994). In this case, as with learning how to choose frames of reference, the way children organize their spatial descriptions is likely to be more a function of their understanding of listener needs than of their spatial representations.

There are several findings to support this analysis. Children of 6, 9, and 11 years have equivalent ability to learn their way through a maze, but the younger children are substantially less skilled in giving coherent verbal directions to someone else (Allen, Kirasic, and Beard 1989). Similarly children as young as 6 years are able to direct listeners to sets of objects in their homes using the descending hierarchical strategy, sending the searchers first to a larger region and then focusing them on progressively smaller regions and landmarks near the target object, but they do not do this spontaneously (Plumert et al. 1994). Six-year-olds produce this kind of organization only when the adult listener reminded the children to give her directions "so I don't have to walk very far."[1] Children of 8 years, but not 6 years, show spatial organiza-

---

1. Children did better when asked to deal with only a single location at a time, however, suggesting that taking listener's needs into account may partly be delayed because it requires information-processing capacity.

tion when planning a tour of a dollhouse, as opposed to simply free recalling the objects in the dollhouse (Plumert and Strahan, 1997). Similarly 8-year-olds, asked to describe a funhouse they had gone through, give a mental tour, whether or not they were told to think in terms of a route through the funhouse, whereas 6-year-olds give a mental tour only when instructed to notice the route (Gauvain and Rogoff 1989). Thus it appears that, by 6 years at least, children have well-organized spatial representations as well as linguistic competence; what they lack, and what they develop between 6 and 8 years, is the ability to analyze what listeners will need to know in order to be able to follow an efficient route through a space.

Summary
Two of the three issues that people need to address in order to talk about space turn out, developmentally, to have more to do with children's understanding of communicative negotiation than with their spatial understanding. By the time children begin to talk about space, they code spatial location in relation to both their own position and in relation to external referents. Thus they have the spatial knowledge required to use linguistic frames of reference, even though it may take time for them to understand that what they think is natural to talk about may be ambiguous to their listeners. Similarly they have route and survey knowledge sufficient to structure sequential descriptions of spatial layouts. It is not, however, until middle childhood that they appear to realize that an organized strategy of spatial description is essential to listener comprehension. Such results are similar to those obtained in many studies of referential communication, in nonspatial as well as spatial domains. Children in the early school years are simply not efficient at recognizing communication failures, although they can do so when given explicit training (Sonnenschein and Whitehurst 1984). Presumably, in the course of normal development, feedback from confused listeners and/or from mishaps resulting from ambiguous communication drive the development of organized description strategies and explicit marking of frames of reference.

By contrast, the other issue involved in talking about space, namely the way in which languages carve up continuous space into regions referred to by spatial words, poses unsolved developmental questions. It is not yet known to what extent children simply map words onto preexisting spatial concepts, as opposed to using the distributional evidence of spatial input language to guide their exploration of spatial distinctions only existing in an inchoate form prior to their hearing adult speech. This issue has general implications, extending beyond the understanding of language acquisition. We have already written, in chapter 4, about the evolution of spatial categories in the course of development. There we attributed developmental change to increased capacity to deal with several dimensions at once and to the feedback that children get regarding their success or failure at finding objects, using more or less finely divided spatial areas. However, it is possible that the spatial distinctions made in a language also influence the evolution of spatial categorization.

It is interesting to contrast the acquisition of spatial communication with the acquisition of the ability to use maps and models to communicate spatial information. We have emphasized that language poses challenges to the communication of spatial information and that maps and models seem better suited, intrinsically, to the expression of multiple metric spatial relations simultaneously available for visual inspection. Despite these facts, however, our impression is that children learn to talk about spatial information more easily than they learn to use graphic forms, and they require less environmental scaffolding in order to do so. The fact that some of the component skills for spatial communication, namely language and understanding social interaction, are species-typical adaptations may be relevant here. Map and model use may be later additions to the human repertoire, like the use of visual symbols to stand for speech. Both reading and map/model use may be dependent on basic human skills rather than more directly supported. In such cases carefully provided environmental support may be especially important to ensure optimum development and to prevent particular individuals from lagging behind.

Chapter 8

Thinking about Development

In this book we have discussed the current state of research and theory concerning development in the spatial domain. Having done so, we are in a position to reflect on what we have learned: not only about spatial development in particular, but also about cognitive development in general. Thinking about cognitive development has gone through several phases. Piaget's account of development dominated for some time, but research eventually led to doubts about the theory, especially about certain aspects of it. It is now generally agreed that Piaget underestimated the richness of the starting points available at birth to the developing child, and that he advocated an overly domain-general account of development with too heavy an emphasis on qualitatively distinct stages. A period followed during which difficulties with these elements of Piaget's thinking were taken to undermine the general project of interactionism. During this time, various forms of nativism became popular and even dominant accounts of cognitive development, although other approaches, such as Vygotskyan theories, were explored as well.

More recently, however, new interactionist accounts of cognitive development have emerged. These accounts postulate richer starting points than Piaget did and have tended to use a domain-specific style of analysis. However, they maintain important elements of Piaget's meta-theory. An initial step in this re-evaluation of the possibilities of interactionism was Karmiloff-Smith's (1992) book, in which she argued that the idea of domain-specific analysis and rich starting points did not presuppose the existence of innate modules. More recently variability-and-selection approaches to cognitive development (Siegler 1996), certain kinds of connectionist (or adaptive systems) models (Elman et al. 1996), recurrent neural network theory (Edelman 1992), and "theory theory" approaches (Gopnik and Meltzoff 1997) have all taken a broadly similar point of view, in which biologically specified starting points interact with environmental input in a way that allows for the emergence of new forms from old ingredients. Meltzoff and Moore (1999) have also argued that strong starting points are compatible with conceptual change, offering a specific analysis of change in infancy as an example.

In the initial half of this chapter, we consider the account of spatial development in this book as an example of this new style of theorizing. There are several themes in this discussion. First, some of the developments we have charted can be understood as the interaction of biological starting points with universally available aspects of the environment. These aspects of the environment, such as the fact that objects are solid or that humans move, are unlikely to differ much (usually not at all) across individuals. Developments in areas dependent on such aspects of the environment are invariably observed, but not because they are innately given. Second, other developments, notably in areas of symbolic spatial competence and spatial reasoning, involve aspects of the environment that show much more marked variability across individuals, due to the relevance of cultural transmission and specialized experiential opportunities. Developments in these areas, accordingly, are much more variable in pace and even in nature across children and cultures. Third, in the course of this discussion of universal and variable aspects of the environment, we consider various cases in which it has been claimed that maturation of relevant brain areas leads to new kinds of behavior. We point out that the claim may be true, but that the conclusion has not yet been fully justified.

In the second half of the chapter, we turn from the spatial domain to consider commonalities in development across other cognitive domains, including quantitative ability, theory of mind, and language. Some of these commonalities revolve around the simple point that interactionist accounts of all three of these domains are gaining in credibility, despite the fact that strong nativist and modularist claims have been made about each of them. In addition there are some specific cross-domain points of contact, for instance, in the role of coding of continuous quantity in both the spatial and the quantitative domains. Both kinds of commonalities give us the hope that domain specificity in cognition can be honored at the same time that, at a more abstract level, a general theory of cognitive development can be constructed.

Spatial Development

Starting Points and Experience-Expectant Interactions

The idea of "environmental influence" is bound up for many people with a conceptualization of the environment as something that is variable across individuals, families, and cultures. While such variation is certainly marked in some cases (e.g., teaching practices, the availability of maps and other cultural artifacts), in other cases environmental input is universally available and without any real variability. That is, there are aspects of the physical and social environments that are always present (e.g., objects are governed by the laws of motion) or almost always present (e.g., when an object moves, its color remains constant). When interactions between biology and the environment involve the environment providing such universally available information, the importance of the environment is often overlooked or minimized. One reason for such omission may simply be that the relevant aspects of the environment are so obvious to us as adults that it seem unnecessary to mention them or assign them theoretical importance. And yet, if infants sometimes emerged into a strangely altered world (e.g., with zero gravity), their perceptual and cognitive development might well be profoundly different.

Controversy often occurs in thinking about development in domains in which environmental input is an inevitable aspect of living in the world. In such domains nativists argue that the environment is merely a "trigger" for processes that are set to unfold by biological endowment. The criteria for distinguishing a trigger, however, are not clear. For instance, when experience with locomotion changes infants' spatial coding practices, is locomotion and the attendant experience a trigger or an experience-expectant interaction? Does it matter just how much experience with locomotion is required, and if so, is the criterion for a "trigger" less than an hour or a day, a week, or a month? A better way to consider such developmental interactions is as experience expectant. This term more properly acknowledges that the environment plays a crucial, even if normally inevitable and ubiquitous, role in producing mature competence.

In the spatial domain there are several important capabilities that infants have when they are born or that they quickly develop in the first few months. These early capabilities change, however, as the infant (and then the child) interacts with the spatial world, keeping track of the outcomes of behaviors and noting regularities. Many important changes occur over the first year, in the course of initial calibration to the world, but other changes continue over the next nine years or so as the system of location coding is fine-tuned through use. The developmental changes that occur are largely regular and universal, although perhaps varying slightly in pace as a function of factors such as motor development. In each of the three examples we mention, we see examples of biology interacting with universal aspects of the spatial environment. There may also be maturational aspects to some of the changes, although determining whether this is true poses empirical challenges not yet well addressed.

Spatial Coding Systems

Very young infants have a nervous system capable of registering visual, auditory, kinaesthetic, vestibular, and proprioceptive information in a spatially differentiated fashion, that is, of encoding input in these modalities in retinal and body-centered coordinates. They are capable of performing computations involving this information, as when retinal coordinates are changed into body-centered ones. They are also capable of using this computed information to guide further exploration of the environment. These abilities allow for the early availability of three systems of spatial coding: cue learning, response learning, and dead reckoning. They allow, in addition, for coding the location of objects in continuous space, an ability necessary for the development of place learning (the fourth fundamental spatial coding system).

Infants also come equipped with the ability to store information on frequency and recency of responses in various systems, and the outcomes of these responses. The failure to locate an object in various kinds of infant search serves as an obvious and direct source of information to the organizing system, and as a signal to experiment with the success of initially less-favored coding systems. Infants develop by determining which of these kinds of information has the highest cue validity, based on their experience. Specifically, during the first year of life, infants begin by relying to a great extent on response learning for spatial localization. Since they are largely stationary, such reliance is not a bad thing—for a stationary being, both response learning and cue learning lead to the same (correct) response, and dead reckoning is irrelevant. But given movement, response learning is unreliable. Adaptive functioning now requires that infants learn that when response learning fails to locate an object, formerly little-used coding systems (i.e., cue learning and dead reckoning) lead to the correct choice. Similarly, in A-not-B search paradigms, a shift occurs from directing reaches based on the most frequent and most recently executed motor response to a reliance on the most recently executed looking response.

While response learning and cue learning are simple systems, not changing much if at all with development except in the occasions of their use, dead reckoning and place learning continue to change for years. Fine-tuning of the dead reckoning system continues over early childhood, likely based on continued use of information from sensory flow, proprioception, and kinaesthesis. Calibration of this system allows for both better determination of position when walking without vision, and the availability of a method of determining location complementary to (and usually mutually supportive of) an external-referent system. The place learning system is not apparent, so far as we know, before the age of 21 months, and changes in its precision and range of use continue until the age of 7 years.

Discussing place learning and the A-not-B error in the context of interactions of biological starting points with universally available input raises the important issue of what role maturation of biological systems may play in this interaction. Arguments have been made for the importance of maturation of dorsolateral prefrontal cortex in the ability of infants to resist the A-not-B error, and the appearance of place learning could depend on maturation of hippocampal areas known to support such relational learning. However, the difficulty with maturational arguments is that they are often made without considering the possible role of experiential factors. For instance, looking back and forth between target objects and distal landmarks is essential for place learning and may only occur after upright locomotion has been achieved. Hippocampal areas might be recruited for such coding, were it to occur; that is, they might be mature and awaiting such input prior to 21 months. Or, they might be immature but await the relevant input in order to achieve their mature form, perhaps through some form of pruning or synaptic selection. The fact that development of place learning continues into early childhood, with mature levels on place tasks not being seen until about 7 years of life, suggests the relevance of accumulating experience.

At present, there is little we can say about whether or not there is such a thing as ongoing nervous-system maturation that occurs independently of environmental experience. The reason is that such a process has often been assumed without any consideration of the role of environmental input. Experimental work with deprivation and enrichment conditions, as well as with naturally occurring variation, may allow us to analyze the nature of maturation more thoroughly in the future. For instance, place learning could be examined as a function of age at walking and when walking is sufficiently independent that looking around rather than down or straight ahead becomes common.

Hierarchical Coding
Hierarchical combination across levels of coding has been seen as early as 16 months. Its origins are not yet known. Both fine-grained and categorical coding have been observed, in separate programs of research, by 6 months. So, combination might be a basic process, present as soon as these two basic ingredients are evident. However, it is also possible that some amount of experience in interacting with the physical and social environment is necessary in order to induce combination and adaptive weighting of different grains of spatial information.

In either case the existence of early hierarchical coding by 16 months does not imply that there is no developmental change in hierarchical coding. Actually there are two lines of development. First, over a period extending through 10 years, children begin to subdivide perceptually given areas so that their spatial categories encompass smaller areas than those used by younger children, and are independent of the existence of perceptually evident boundaries. Such developmental change means that the advantages of hierarchical combination for better overall accuracy in judgment are increasingly evident, since smaller categories allow for correction of fine-grained coding with smaller amounts of bias. Second, there is substantial developmental change, over a period extending through 10 years, in the ability to conduct hierarchical combination along two dimensions at once. Younger children can hierarchically combine information from one dimension at a time, but when there are two dimensions, they appear to use hierarchical combination along only one of them. This fact makes them less accurate in judgment overall than older children.

Both of these lines of development could plausibly be based on continued experience with the success and failure of localizing efforts under variable methods of coding, that is, depend on a variation-and-selection mechanism. However, there could also be a role for maturational processes. For instance, being able to hierarchically code along more than one dimension at a time could depend on increases in working memory capacity. However, as we said before, sufficient work has not been done to allow for bold maturational claims. In the case of working memory capacity, doubts about maturation are particularly strong because there is an obvious way in which working memory capacity could increase nonmaturationally, based on increasing knowledge about the world and consequent chunking of material in working memory.

What Is an Object?
Initially, infants seem to use spatiotemporal information as a primary means of identifying objects—an object is defined by where it is. By 12 months, they come to define objects as adults do, linking the experience of particular static perceptual characteristics (e.g., shape, size, and color) to particular spatiotemporal pathways. It remains to be determined whether the first year of life sees a gradual shift in the use of various attributes as defining objects, dependent on the information-processing and problem-solving capacities of infants, or a more sudden and qualitative shift. In either case the precipitating mechanisms for the insight need to be identified. One possibility is that simple Humean induction leads to the expectation that perceptual characteristics such

as shape, size, and color are unchanged if a spatiotemporal path is continuous. A very different alternative is that the highlighting of continuity of static perceptual characteristics through linguistic labeling leads to the shift. And, of course, there is always the possibility of a maturational change at this age that would lead to the acquisition of a mature object identification system. We raise this last possibility only to note, one more time, that investigation of maturation is an important issue that will require intensive investigation in the future.

Variable Environmental Input

In discussing environmental input so far, we have concentrated on input that is universally available given the very nature of the spatial world, such as, the fact that objects must occupy space and must necessarily have a location with respect to each other and to an observer. However, other aspects of spatial competence, such as spatial reasoning, spatial language, and map use, while resting on a foundation of simpler skills, also require abilities and strategies that are almost certainly more dependent on cultural transmission, with input likely to be more variable across children and social groups. For some domains, such as language development, it is clear at least in principle how to examine the problem of input (i.e., analyze collections of adult speech to children). For spatial reasoning and symbol use, it is not as self-evident what the relevant input is, although one can guess that exposure to spatial problems, maps, spatial language, or the need to navigate independently might be important. There is a great need for methodological advances to allow the analogous study of the nature of input in the other domains of cognitive development, including the spatial domain.

Development of Spatial Reasoning
Probably the best-known developmental transition in spatial reasoning in later childhood is the acquisition of the ability to look at pictures of a scene viewed from different vantage points and identify which one corresponds to the view of a person at a designated point. This task reflects the ability to deal with two conflicting frames of reference, the physically present one and the rotated one viewed from another perspective. Using the latter, as required for picture selection, is difficult, and children are 9 or 10 years of age before they are successful. However, children as young as 3 years can succeed in answering questions about other points of view if asked questions that do not require dealing with a conflict of frames of reference, such as, "If you were over there, what would be on your left side?"

How does the ability to deal with conflicting frames of reference emerge for tasks such as picture selection that require such an ability? Change might be guided by success and failure experiences in accurately predicting transformed scenes. A switch to use of an imagined frame of reference could also be aided by use of verbalizable strategies, in particular, the strategy of dealing with only one object in the array rather than the whole scene. Discovery of such strategies might be a self-generated insight, or it might arise from instruction or observation of others. Unfortunately, there is no direct evidence regarding what kinds of experiences children actually have with such problems in everyday life or what kinds of feedback they obtain. This is one area where it is at least reasonably evident how to go about gathering relevant input evidence, since instances of perspective taking are well defined. However, a practical difficulty with such research may be that the input is relatively sparse, so getting good observational measures would be costly.

Talking about Space

Children's talk about space begins at an age when they have already acquired much of the spatial knowledge required to use linguistic spatial terms. Thus the foundational abilities for learning to talk about space are the early location codings that we have already discussed, built up through the interaction of biological staring points with universal input. However, exposure to spatial terms also appears to heighten or sharpen children's attention to potential spatial distinctions (an idea that can be termed the accessibility hypothesis). Children have a variety of concepts available—they can observe, for instance, the difference between tight-fitting or loose-fitting containment—but the consistency and frequency of use of such an available distinction affects the position of that distinction in a salience hierarchy. Alternatively, or in addition, it may be that exposure to spatial terms initiates a search for relevant spatial distinctions that such terms encode. Either way, the linguistic environment in interaction with the observing and hypothesis-testing child produces a mature system of spatial reference. Children's acquisition of such a mature reference system has effects in turn on further spatial development, as children become able to discuss spatial concepts (e.g., length) in a fashion that communicates precisely with other members of their linguistic community.

Development of linguistic communication of spatial information also depends on children's understanding of their listeners' comprehension needs. For instance, children are in elementary school before they realize that an organized strategy of spatial description is essential to getting listeners to understand the description. Feedback from confused listeners and/or from mishaps resulting from ambiguous communication may drive development in this area.

Development of Use of Maps and Models
Children begin to be able to use simple maps and models in simple situations at about the same time they show evidence of spatial reasoning, that is, before the age of four years. In other words, they can use simple maps presented to them in alignment with the real world to learn about the location of objects denoted by arbitrary visual elements. In addition they have some ability to learn about distances from such simple maps and to coordinate oblique and eye-level views of a space.

The foundational abilities for the development of map and model use may be two achievements. One foundation is "representational insight," the idea that something can stand for something else, which begins in a humble form toward the end of the second year but takes another year or two to be applied to the variety of situations in which human beings use it. Thus this foundation for map use is itself constructed over the course of several years, in the various kinds of interaction with symbolic representations and the people who interact with them. The other foundation for map use is relative spatial coding, that is, thinking of an extent as a certain part of a perceptually present referent (e.g., a quarter of the way across). Such relative coding allows for the use of scale in simple situations. The origins of this building block are not well known, although there is some suggestion that it too comes as an insight—children seem either quite good at using relative coding to transform scale or to not to be able to perform the task at all.

Despite these strong beginnings sometime in the fourth year, children's early map use is hardly mature competence. Map- and model-reading skills greatly improve over the rest of the first decade. Interpretation of scale becomes increasingly more precise as measurement and calibration skills grow and children maintain scaling over multiple relations. Alignment between map and world becomes easier (but never simple) to establish, as strategies for alignment are found and the efficiency of mental rotation improves. Planning and comparison skills emerge which allow for the use of maps in multistep navigation tasks.

It is likely that environmental input is crucial to these developments, since the particularities of mapping conventions are culturally relative and must be taught while the capacity to symbolize is likely something for which humans have been evolutionarily selected. The fact that certain spatial skills akin to mapping improve more during the school year than during the summer provides some support for this assertion.

But exactly what kind of instruction is important and what may occur during early years at home or in preschool is not yet known.

Commonalities and Differences with Other Domains

Theorizing about other domains, including the development of quantitative understanding, the origins of a theory of mind, and the acquisition of language, has followed a somewhat similar course to thinking about spatial development. For each domain, there are approaches that emphasize the role of innate endowments in shaping development, approaches that emphasize the role of environment and culturally transmitted tools, and interactionist approaches that often began as proposals from Piaget. Recent thinking and studies in each domain have led to the proposal of new, non-Piagetian, interactionist views that integrate perspectives. Typically, early competence is delineated in a way that balances examination of starting points with an examination of important developmental transformations and how they depend on experience and culture.

Quantitative Development

Theorizing about quantitative development and theorizing about spatial development have had quite similar histories. Initially, in the quantitative domain as in the spatial domain, Piaget's views (Piaget 1941/1965) predominated, and it was widely believed that children could not reason quantitatively until the early elementary school years. New techniques, however, led to the demonstration of quantitative abilities at younger ages, initially in preschoolers (e.g., Gelman and Gallistel 1978), and later in young infants (e.g., Starkey and Cooper 1980; Starkey, Spelke, and Gelman 1990; Wynn 1992). For instance, a particularly striking experiment was Starkey et al.'s (1990) finding that young infants showed evidence of cross-modal matching of number, between visual representations of certain numbers and auditory sequences consisting of the same number of sounds. These data seemed to indicate the presence of a sense of number that was abstract, given its cross-modal nature. These and other findings of early competence led to strong nativist claims about the bases of human quantitative ability (e.g., Starkey 1992). On the other hand, work pointing to the importance of cultural contexts and symbolic systems in quantitative development (e.g., Miller and Stigler 1987; Saxe, Guberman, and Gearhart 1987) supported interest in a Vygotskyan perspective.

Recent evidence calls the nativist viewpoint on quantitative development into question. An initial departure point was Huttenlocher, Levine, and Jordan's (1994) investigation of the origins of the ability to solve nonverbal calculation problems. These investigators had already shown that by 4 years, children who (not surprisingly at this age) still have difficulty with conventional arithmetic presented as number-fact problems or story problems can solve simple addition and subtraction problems presented in a nonverbal format (Levine, Jordan, and Huttenlocher 1992). In a nonverbal format, an examiner shows the child a certain number of disks, moves those disks behind a screen, and then adds more disks to the array behind the screen; the child is asked to construct an array of disks showing the number that is behind the screen. Huttenlocher et al. (1994) found that this ability to solve nonverbal calculation tasks is not present from the outset. It emerges only as children approach 3 years of age, is initially approximate and is highly related to overall intellectual competence. Such findings do not easily fit notions of innate numerical competence.

Further doubts about a nativist approach arise from a study by Mix, Huttenlocher, and Levine (1996). This paper found that 3-year-old children perform at chance on an auditory-visual equivalence test. This is an odd deficiency to find in preschoolers if infants can appreciate such equivalence, as claimed by investigators such as Starkey. When children in their fourth year of life have difficulty with tasks on which neonates have supposedly shown competence, one wonders about the nature of that early competence. (Invoking U-shaped developmental curves is an explanation of the pattern of the data only if there is an associated account of what leads to the waning and waxing of the ability.)

Subsequently, direct evidence has emerged suggesting reason to be skeptical that infant competence with quantity is as great as the initial round of studies seemed to show. A study by Mix, Levine, and Huttenlocher (1997) found that infants did not seem to have the ability to notice quantitative equivalence between auditory sequences and visual displays initially suggested in the Starkey work. Other recent data suggest that infants in "number" experiments may be responding to overall contour length, area, or mass rather than discrete number (Clearfield and Mix 1999; Feigenson and Spelke 1998). Thus, taken together with the earlier study of preschoolers, there is little reason to think that infants have the ability to deal with cross-modal quantitative correspondence or even discrete quantity (i.e., number).

This body of work suggests a new way to think about development in the quantitative domain. There are at least three proposals regarding initial departure points for quantitative development. One idea, based on Meck and Church's (1983) accumulator model, is that infants can form inexact representations of discrete number. Such an ability could be based on a simple neural mechanism, since it is exhibited even by animals. In this case the emergence of nonverbal calculation abilities would be linked either to increasing exactness of the accumulator mechanism (i.e., narrower error distributions around the correct response) or to the replacement of the accumulator by a truly exact mechanism. Such a new mechanism could be based on learning the count words and mapping them to the central tendencies generated by the accumulator. The second possibility is that infants begin by representing quantity with an approximate mechanism based on continuous amount rather than discrete number (Huttenlocher 1993), as supported by recent data of Clearfield and Mix (1999) and Feigenson and Spelke (1998). Development might then be based in part on increasing efficiency in approximation, but there would also be a need at some point to make a transition to representing and reasoning about discrete amount. A third possibility is that infants (and at least some primates) reason initially about small numbers by opening "object files" each time they see a new object (Hauser and Carey 1998; Simon 1997). In this case the emergence of nonverbal calculation ability might be linked to the emergence of the ability to construct, maintain, and manipulate mental models of objects and their motion (Huttenlocher et al. 1994).

Although these accounts differ in important ways, for the present purposes it is more interesting to see how they agree. On all three hypotheses there are crucial developmental transitions moving the infant from initially limited representations to the ability to acquire exact knowledge of quantity, number, and the effects of movement of items into or away from an array of objects (i.e., simple addition and subtraction). These transitions could well be based on interactions of biological starting points with environmental experiences, although, as with aspects of spatial development, it is possible that there are maturational components as well.

Nor need any of the three approaches neglect the role of the cultural environment. Either as a crucial part of the transition to true number ability, or subsequent to it, children learn to use language and other symbol systems to refer to number concepts. As reviewed by Dehaene (1997), there have been significant inventions in our ability to symbolize number, notably the invention of Arabic numerals including a symbol for the essential concept of zero. Without such a symbol system, simple operations of adding, subtracting, multiplying and dividing are extremely difficult. In addition cross-cultural differences in how numbers are given names (long or short; logically structured or idiosyncratic) seem to affect the pace of development and even the rapidity of adult arithmetical processing. Probably any symbol system is better than none at all, but if children are lucky enough to be born into a culture with an efficient notational system, they may be helped considerably in dealing with larger numbers and arithmetical operations. Given the social nature of symbol systems, it is not surprising (although it is distressing) that conventionalized arithmetical abilities, unlike nonverbal arithmetic, are strongly linked to social class and opportunities to master culturally transmitted tools (Jordan, Huttenlocher, and Levine 1992).

This account of early quantitative development has some points of contact with our account of spatial development, both at a specific and at a general level. For instance, in terms of specifics, the accounts share the hypothesis that the appearance of representational ability, or the ability to construct and manipulate mental models, may be an important transition in early development. Even more specifically, if initial representations of number are based on continuous amount, these representations may be quite similar to those that support initial spatial codings. That is, infants may pay attention to the amount of stuff (primitive numerical quantity) and the amount of space (primitive distance) in much the same way. Huttenlocher and Gao (1998) have in fact found that infants are sensitive to the amount of liquid in a container and to changes in that amount, just as Newcombe et al. (in press) found that infants were sensitive to where an object had been buried in a sandbox.

In general terms, both accounts are broadly interactionist, simultaneously respecting and interweaving consideration of initial starting points, biological changes, interactions with the physical environment, and the influence of culturally transmitted tools. For early quantitative development, there has been even less work than for early spatial development relevant to the issue of what experiences may lead to change in the system, namely, what mechanisms might shape the formation of discrete mental models out of more continuous representations, or might lead to the recognition of cross-modal numerical correspondences. For later development in the quantitative domain, there has been much more work than in the spatial domain on the mechanisms of developmental change, perhaps because the various strategies for approaching specific arithmetical tasks have been so clearly delineated. For instance, developmental change in the use of various techniques for finding the answer to simple addition problems has been modeled in terms of adaptive strategy choice (Siegler and Shipley 1995).

There are contrasts between the quantitative and spatial domains as well as similarities. One important difference is that conventional, culturally transmitted systems for conceptualization and communication appear far more important for development in the quantitative than in the spatial domain. While maps, diagrams, navigational systems, and spatial language are invaluable tools, that greatly augment human spatial ability, we can find our way around the spatial world without such aids. By contrast, there is reason to suspect that without count words and numerical symbol systems, quantitative reasoning might remain approximate and inexact. At least on some theories of quantitative development, the very idea of number emerges after several years of development, and it is based at least in part on learning discrete numerical symbols. Coding spatial extent and spatial relations are perhaps more basic attributes of the human nervous system.

Theory of Mind

There have been some of the same distinct lines of thought about theory of mind (i.e., the ability to understand other people's thoughts and feelings) as about the spatial and quantitative domains. However, they have not been entertained in the same order of predominance. While Piagetian theory was initially the leading way to think about spatial and quantitative development, research on theory of mind began when nativist thinking about cognitive development was already popular, and such a framework was applied from the start to theorizing about children learning about other people's thoughts and emotions (e.g., Baron-Cohen 1995; Fodor 1992; Leslie 1992, 1994). Theory of mind has been widely postulated to be an innate and encapsulated modular ability.

Interestingly this claim has been made despite the fact that successful performance on many tasks said to index theory of mind does not emerge until between the ages of 3 and 4 years. Nativists get around this difficulty in one of two ways. Fodor (1992) suggests that apparent changes in the ability to predict others' performance are illusory—very young children have the same competence as adults but are simply limited from displaying this competence by the lack of a few heuristics. By contrast, Leslie (1992, 1994) offers a maturational account in which transitions in children's performance occur as portions of the theory of mind module come "on line." Similarly Baron-Cohen (1995) suggests that two modules relevant to theory of mind, a module for detecting direction of eye gaze and a module for detecting intentionality, are innate. He thinks that a third module develops later, from the conjoining of the innate modules, and a fourth module, the full-blown theory of mind module, comes from the ability to represent and reason about the contents of the third module.

A prominent piece of evidence for a modular approach has been the finding that children with autism can display specific deficits in theory of mind tasks (e.g., Baron-Cohen 1995, ch. 5). Their problems are sometimes contrasted with the cognitive profile of individuals with

Williams syndrome or Down syndrome, who are said to have excellent social intelligence while suffering from different specific deficits in other cognitive abilities. However, the claim that children with autism have a specific deficit in the area of theory of mind has recently been questioned. Yirmiya et al. (1998) conducted meta-analyses showing that although individuals with autism had more severe theory of mind deficits than individuals with mental retardation not due to Down syndrome, the latter group was still significantly impaired on theory of mind. Thus a theory of mind deficit is not specific to autism, nor uniquely associated with autistic symptoms. A similar kind of argument can be made on the basis of recent findings that children with deafness also show theory of mind deficits, although they are by no stretch of the imagination autistic (De Villiers 1999). In addition recent evidence has questioned Baron-Cohen's model of how a module for detecting eye gaze might operate and develop (Lee et al. 1998). The development of the capacity for inferring another person's state of mind from eye gaze seems to involve a gradual developmental process in which a large number of contextual cues need to be considered and integrated to make a plausible judgment.

The innate-modular approach to theory of mind has come under critical attack on other grounds as well (Moore 1996). Moore presents evidence that the argument for localized areas in the prefrontal areas of the brain dedicated to a theory of mind module is weak and contradictory. Even if such localization existed in adults, it might emerge during development rather than be present from the start. Children with autism, who are said to have specific theory of mind deficits, may have differences in other cortical and subcortical areas, instead of or in addition to prefrontal areas (see the review by Waterhouse, Fein, and Modahl 1996).

Perhaps most important, Moore argues, there is no necessary connection between evolutionary psychology and a nativist-modularist theory of development despite enthusiastic claims to the contrary (e.g., as in Cosmides and Tooby's introduction to the Baron-Cohen book). The Cosmides-Tooby point of view is that the brain has evolved in the course of natural selection to solve adaptive problems (an indisputable truism) but that, further, it has done this by evolving multiple modules, each designed to address a particular adaptive problem. However, as Moore points out, it is not clear exactly what adaptive problem is solved by a theory of mind module. Social development does not have a narrow range of application, as a modularist would suggest, but rather is relevant to a number of adaptive functions, as different as mating and feeding. A planning mechanism that takes other minds into account is useful to a social animal and likely does result from adaptive pressure. However, such a mechanism can as well be something that is emergent in the course of development, given certain biological starting points, as specified in detail from the start.

There are many nonnativist counterproposals about development of a theory of mind. One of these approaches is a "theory theory" (Gopnik and Wellman 1992, 1994). Gopnik and Wellman propose that children initially understand other people only in terms of their desires or wishes; that is, people do what they want to do. But predictions made on the basis of this desire theory sometimes result in puzzling failures. As such failures accumulate, by around the third birthday, children develop a second theory, which postulates thoughts and beliefs as well as desires and wishes. Initially, however, thoughts and beliefs are not central to children's thinking but are only invoked when a desire-belief understanding fails. This theory is better than the first, but it still does not effectively predict what other people do in certain situations. Finally, by around 4 or 5 years of age, children have evolved a third theory, in which representations of others' beliefs play a central role.

The Gopnik and Wellman theory theory is not the only nonnativist proposal for understanding development of theory of mind; there are in fact several possibilities (e.g., Frye, Zelazo, and Palfai 1995; Harris 1994; Perner 1991). For the present purpose, the important message is that the competing proposals are all fundamentally interactionist and involve solving the adaptive problems facing a social animal by postulating abilities that are almost certain to develop given interaction of biological starting points with universal aspects of the environment. In each account infants begin the task of acquisition of a domain with certain key skills, perhaps including such components as an eye-detection detector. However, these starting points are not regarded as the essence of adult competence but are rather seen as transformed in the course of environmental transactions. The nature of these transformations is in dispute, as are the nature of the essential experiences and whether domain-general abilities such as information-processing capacity are relevant to the domain-specific tale. Still these crucial differences should not obscure the important commonality, which is a framework neither purely nativist nor purely environmentalist.

Language

Language is a domain somewhat different from the other domains we have talked about so far in that it provides a system for symbolizing and communicating about other domains (e.g., space, number, and other people's intentions). Yet it has been regarded as a preeminent example of a domain (e.g., Fodor 1983), and the dominant view of its acquisition has been a nativist one, beginning with Chomsky (1965) and continuing today (e.g., Pinker 1994). There have, however, been vigorous dissents to nativism in the work of investigators who have taken the alternative view that general cognitive abilities and cognitive development allow for the explanation of language acquisition, a point of view originally advocated by Piaget but considerably augmented and rethought by modern theorists (e.g., Bates and MacWhinney 1989). Other investigators have argued the essentially Vygotskyan point that adult input and social interaction play a key role in language development (e.g., Nelson 1985, 1996; Tomasello and Farrar 1986).

Recently, however, as with space and quantity, more integrative interactionist viewpoints have been developed. Discussion of language acquisition is, for example, a central element in the Elman et al. (1996) argument for a more thoroughly interactionist view of development. Elman et al. are proponents of a connectionist approach to language acquisition, in specific, but they also argue, in more general support for an interactionist point of view, that many of the phenomena cited by nativists to bolster their case have quite natural nonnativist explanations. For instance, children suffering from specific language impairments may not have a truly language-specific problem. Their problems with language may instead begin with a difficulty with processing fast auditory transitions, and they may also suffer allied, but more subtle, deficits in areas such as balance. Similarly children with Williams syndrome may not have a uniquely spared and intact capacity for language and face processing. Neither their grammar or their syntax is completely normal; more important, their performance may be based on different cognitive processes than the performance of typically developing children (Karmiloff-Smith 1998).

Another notable example of interactionist thinking about language acquisition is the joint work of Hirsh-Pasek and Golinkoff, who argue for a "coalition model" of the development of grammatical comprehension (Hirsh-Pasek and Golinkoff 1996) and of early word learning (Golinkoff, Hirsh-Pasek, and Hollich, 1999; Hollich, Hirsh-Pasek, and Golinkoff, under review; see also Deák, in press). In this way of thinking, infants begin the task of learning language, both vocabulary and syntax, with certain initial biases. For example, they focus at first on prosodic information in speech. This concentration is adaptive because it allows for the segmentation of the speech stream into phrases and clauses. Such a concentration also allows for initial hypotheses about the relation between acoustically defined sound units and possible meanings. As another example, infants begin the task of learning words with a bias to focus on the perceptual salience of objects as the determiner of what a novel sound refers to. These initial biases are just starting points that change developmentally, allowing for different mechanisms and strategies to be brought to bear on language acquisition. Crucially, for Hirsh-Pasek and Golinkoff, children are exposed to multiple inputs at any given point in time, but these inputs are differentially weighted in the course of development. In acquisition of syntax, for example, Hirsh-Pasek and Golinkoff suggest that toward the end of the first year, infants' biases in learning shift away from prosody so that semantic cues become relatively more important. In word learning, during the second year, infants come to be guided less by perceptual salience and more by social and contextual cues. In the end, children are in a position to work on abstract grammatical relations in syntax and to be guided in word learning by constraints and principles, such as a search for whole objects to which a novel word could possible refer.

As with our approach to spatial development or new approaches to quantitative development, we see in these new approaches to language acquisition the delineation of starting points for development that are neurally plausible and domain specific. These starting points are more than what Piaget was willing to give the infant but less content-laden than the assumptions of classic nativists. We see also a developmental perspective in which key transformations occur as learning progresses, and an integration of the cultural and interpersonal environment (the Vygotskyan perspective) with the initial starting points stressed by nativists and with the interaction of a thinking organism with the world emphasized by Piaget.

A Last Word

In summary, interactionist thinking provides an interesting and promising way to conceptualize cognitive change. This family of theoretical approaches is, interestingly, similar in some ways to Piaget's original thinking, most evidently on the level of a meta-theory of development. Both Piaget and many of these recent theories see development as change in an organism simultaneously constrained by biology, physical environment, and social environment. To point out similarities between Piaget and modern developmental theories is not, however, to claim that all these models were anticipated by Piaget, nor that they are equivalent to each other. There are crucial differences from Piaget in treatment of the concept of stages and in thinking about mechanisms of development, as well as numerous local differences in specific descriptions of development. In addition there are certainly crucial differences among the interactionist approaches mentioned in this chapter.

It does seem though that we may now have at least the general outlines for a way to understand cognitive development, not just in the spatial domain but in other domains as well. This would be a welcome change—doctrinal wars have taken up a lot of time in the study of cognitive development. Perhaps they will have been worth the time if they lead to understanding that one can work within a developmental theory that respects (and, more important, specifies) the neonatal starting points for cognitive development, biological maturation, the knowledge gained from individuals' interactions with average expectable environments, and the contributions of social interactions that transmit culturally evolved amplifiers of cognition. In short, environmentalism and nativism dissolve into a real interactionism as one achieves fine-grained domain-specific understanding of development in both phylogeny and ontogeny.

The danger of interactionist thinking is unreflective eclecticism. Development is sometimes portrayed as an unspecified interplay of a list of factors, a theory that is trite and unilluminating. The constraint on eclecticism, and the key to a rich theory, is close observation and domain specificity. In anthropology Clifford Geertz has argued for "thick description" of phenomena as a form of theoretical understanding. The same strategy may pay off for developmental psychology (Kagan 1998). When working on specific problems in a domain such as space (or quantity, or theory of mind, or language), one works at a level in which one formulates testable hypotheses about particular behaviors, as manifested at particular times, in relation to particular kinds of physical and social interactions, and conditioned by particular manifestations of biological maturation. Such an approach does not mean abandoning thought about cross-domain generalities, either at the specific level (e.g., attention to continuous quantity as foundational for both numerical and spatial development) or at a more abstract level (e.g., the importance of variation and selection in development). However, at the present stage of development of our science, there is a strategic advantage to domain specificity and intensive investigation of particular problems. Hypotheses formulated about particular developmental issues cannot be vacuous, precisely because they may be shown to be wrong. This book will have served its purpose if, in a decade, we can say just what hypotheses in it have been confirmed and which have been found wanting.

References

Aadland, J., Beatty, W. W., and Maki, R. H. 1985. Spatial memory for children and adults assessed in the radial maze. *Developmental Psychobiology* 18: 163–72.

Acredolo, L. P. 1977. Developmental changes in the ability to coordinate perspectives of a large-scale space. *Developmental Psychology* 13: 1–8.

Acredolo, L. P. 1978. Development of spatial orientation in infancy. *Developmental Psychology* 14: 224–34.

Acredolo, L. P. 1979. Laboratory versus home: The effect of environment on the 9-month-old infant's choice of spatial reference system. *Developmental Psychology* 15: 666–67.

Acredolo, L. P. 1982. The familiarity factor in spatial research. *New Directions for Child Development* 15: 19–30.

Acredolo, L. P. 1988. From signal to "symbol": The development of landmark knowledge from 9 to 13 months. *British Journal of Developmental Psychology* 6: 369–72.

Acredolo, L. P. 1990. Behavioral approaches to spatial orientation in infancy. In A. Diamond, ed., *The Development and neural bases of higher cognitive functions*. New York: New York Academy of Sciences, pp. 596–607.

Acredolo, L. P., Adams, A. and Goodwyn, S. 1984 The role of self-produced movement and visual tracking in infant spatial orientation. *Journal of Experimental Child Psychology* 38: 312–27.

Acredolo, C., Adams, A., and Schmid, J. 1984. On the understanding of the relationships between speed, duration, and distance. *Child Development* 55: 2151–59.

Acredolo, L. P., and Boulter, L. T. 1984. Effects of hierarchical organization on children's judgements of distance and direction. *Journal of Experimental Child Psychology* 37: 409–25.

Acredolo, L. P., and Evans, D. 1980. Developmental changes in the effects of landmarks on infant spatial behavior. *Developmental Psychology* 16: 312–18.

Acredolo, L. P., Pick, H. L., and Olsen, M. G. 1975. Environmental differentiation and familiarity as determinants of children's memory for spatial location. *Developmental Psychology* 11: 495–501.

Aguiar, A., Kolstad, V., and Baillargeon, R. (in press). Problem solving and perseveration in infancy. In H. Reese, ed., *Advances in Child Development and Behavior*, Vol. 27. New York: Academic Press.

Ahmed, A., and Ruffman, T. 1996. Do infants know when they are searching incorrectly? Looking times in a non-search A not B task. Presented at International Conference on Infant Studies, Providence, RI.

Ahmed, A., and Ruffman, T. 1998. Why do infants make A not B errors in a search task, yet show memory for the location of hidden objects in a nonsearch task? *Developmental Psychology, 34: 441–53.*

Allen, G. L. 1981. A developmental perspective on the effects of "subdividing" macrospatial experience. *Journal of Experimental Psychology: Human Learning and Memory 7:* 120–32.

Allen, G. L., Kirasic, K. C., and Beard, R. L. 1989. Children's expressions of spatial knowledge. *Journal of Experimental Child Psychology* 48: 114–30.

Allen, G. L., Kirasic, K. C., Siegel, A. W., and Herman, J. F. 1979. Developmental issues in cognitive mapping: The selection and utilization of environmental landmarks. *Child Development* 50: 1062–70.

Allen, G. L., and Ondracek, P. J. 1997. Children's acquisition of spatial knowledge from verbal descriptions. Presented at the Biennial Meeting of the Society for Research in Child Development, Washington, DC.

Alyan, S., and McNaughton, B. L. 1999. Hippocampectomized rats are capable of homing by path integration. *Behavioral Neuroscience,* 113: 19–31.

Anooshian, L. J., Hartman, S. R., and Scharf, J. S. 1982. Determinants of young children's search strategies in a large-scale environment. *Developmental Psychology* 18: 608–16.

Anooshian, L. J., and Nelson, S. K. 1987. Children's knowledge of directional relationships within their neighborhood. *Cognitive Development* 2: 113–26.

Anooshian, L. J., and Young, D. 1981. Developmental changes in cognitive maps of a familiar neighborhood. *Child Development* 52: 341–48.

Antell, S. E., and Caron, A. J. 1985. Neonatal perception of spatial relationships. *Infant Behavior and Development* 8: 15–23.

Arditi, A., Holtzman, J. D., and Kosslyn, S. M. 1988. Mental imagery and sensory experience in congenital blindness. *Neuropsychologia* 26: 1–12.

Baenninger, M., and Newcombe, N. 1989. The role of experience in spatial test performance: A meta-analysis. *Sex Roles* 20: 327–44.

Baenninger, M., and Newcombe, N. 1995. Environmental input to the development of sex-related differences in spatial and mathematical ability. *Learning and Individual Differences* 7: 363–79.

Bai, D. L., and Bertenthal, B. I. 1992. Locomotor status and the development of spatial search skills. *Child Development* 63: 215–26.

Bailenson, J. N., Shum, M. S., and Uttal, D. H. 1997. Strategies and asymmetries in route selections on maps. Unpublished manuscript, Northwestern University, Evanston, IL.

Baillargeon, R. 1986. Representing the existence and the location of hidden objects: Object permanence in 6- and 8-month-old infants. *Cognition* 23: 21–41.

Baillargeon, R. 1987. Object permanence in 3.5- and 4.5-month-olds. *Developmental Psychology* 23: 655–64.

Baillargeon, R. 1993. The object concept revisited: New directions in the investigation of infant's physical knowledge. In C. Granrud, ed., *Carnegie Symposium on Cognition: Visual Perception and Cognition in Infancy.* Hillsdale NJ: Lawrence Erlbaum, pp. 265–315.

Baillargeon, R., DeVos, J., and Graber, M. 1989. Location memory in 8-month-old infants in non-search AB task: Further evidence. *Cognitive Development* 4: 345–67.

Baillargeon, R., and Graber, M. 1988. Evidence of location memory in 8-months-old infants in a nonsearch AB task. *Developmental Psychology* 24: 502–11.

Baillargeon, R., Graber, M., DeVos, J. and Black, J. 1990. Why do young infants fail to search for hidden objects? *Cognition* 36: 255–84.

Baillargeon, R., Spelke, E. S., and Wasserman, S. 1985. Object permanence in five-month-old infants. *Cognition* 20: 191–208.

Banich, M. T., and Federmeier, K. D. 1999. Categorical and metric spatial processes distinguished by task demands and practice. *Journal of Cognitive Neuroscience* 11: 153–66.

Barclay, J. R. 1973. The role of comprehension in remembering sentences. *Cognitive Psychology* 4: 229–54.

Baron-Cohen, S. 1995. *Mindblindness: An Essay on Autism and Theory of Mind.* Cambridge: MIT Press.

Bartsch, K., and Wellman, H. M. 1988. Young children's conception of distance. *Developmental Psychology* 24: 532–41.

Bates, E., and MacWhinney, B. 1989. Functionalism and the competition model. In B. MacWhinney and E. Bates, eds., *The Crosslinguistic Study of Sentence Processing.* Cambridge: Cambridge University Press.

Bauer, R. M., and Rubens, A. B. 1985. Agnosia. In K. M. Heilman and E. Valenstein, eds., *Clinical Neuropsychology* 2nd ed. New York: Oxford University Press.

Baylis, G. C., and Driver, J. 1993. Visual attention and objects: Evidence for hierarchical coding of location. *Journal of Experimental Psychology: Human Perception and Performance* 19: 451–70.

Behl-Chadha, G., and Eimas, P. D. 1995. Infant categorization of left-right spatial relations. *British Journal of Developmental Psychology* 13: 69–79.

Bell, M. A. 1998. A looking version of the A-not-B task: Frontal EEG and infant cognitive functioning. Paper presented at a symposium at the International Conferences on Infant Studies Atlanta, GA.

Bell, M. A., and Fox, N. A. 1992. The relations between frontal brain electrical activity and cognitive development during infancy. *Child Development* 63: 1142–63.

Bell, M. A. and Fox, N. A. 1996. Crawling experience is related to changes in cortical organization during infancy: Evidence from EEG coherence. *Developmental Psychobiology* 29: 551–61.

Bell, M. A., and Fox, N. A. 1997. Individual differences in object permanence performance at 8 months: Locomotor experience and brain electrical activity. *Developmental Psychobiology* 31: 287–97.

Bence, V. M., and Presson, C. C. 1997. The spatial basis of young children's use of scale-models. Presented at the Biennial Meeting of the Society for Research in Child Development, Washington, DC.

Benson, J. B., and Uzgiris, I. C. 1985. Effect of self-initiated locomotion on infant search activity. *Developmental Psychology, 21: 923–31.*

Bertenthal, B. I. 1996. Origins and early development of perception, action, and representation. *Annual Review of Psychology* 47: 431–59.

Bertenthal, B. I., and Campos, J. J. 1990. A systems approach to the organizing effects of self-produced locomotion during infancy. In C. Rovee-Collier and L. P. Lipsitt, eds., *Advances in Infancy Research,* vol. 6. Norwood NJ: Ablex Publishing, pp. 1–60.

Bertenthal, B. I., Campos, J., and Barrett, K. 1984. Self-produced locomotion: An organizer of emotional cognitive, and social development in infancy. In R. Emde and R. Harmon, eds., *Continuities and discontinuities in development* (pp. 175–210). New York: Plenum Press.

Bialystok, E. 1989. Children's mental rotations of abstract displays. *Journal of Experimental Child Psychology* 47: 47–71.

Bigelow, A. E. 1996. Blind and sighted children's spatial knowledge of their home environments. *International Journal of Behavioral Development* 19: 797–816.

Bjork, E. L. and Cummings, E. M. 1984. Infant search errors: Stage of concept development or stage of memory development? *Memory and Cognition* 12: 1–19.

Blades, M. 1991. The development of the abilities required to understand spatial representations. In D. M. Mark and A. V. Frank, eds., *Cognitive and Linguistic Aspects of Geographic Space.* Dordrecht: Kluwer Academic Press, pp. 81–115.

Blades, M., and Spencer, C. 1994. The development of children's ability to use spatial relations. In H. W. Reese, ed., *Advances in Child Development and Behavior, vol. 25.* New York: Academic Press, pp. 157–99.

Blaut, J. M. 1991. Natural mapping. *Transactions of the Institute of British Geographers, N.S.* 16: 55–74.

Blaut, J. M. 1997. Piagetian pessimism and the mapping abilities of young children: A rejoinder to Liben and Downs. *Annals of the Association of American Geographers* 87: 168–77.

Blaut, J. M., McCreary, G. S., and Blaut, A. S. 1970. Environmental mapping in young children. *Environment and Behaviour* 2: 335–49.

Blaut, J. M., and Stea, D. 1971. Studies of geographic learning. *Annals of the Association of American Geographers* 61: 387–93.

Blaut, J. M., and Stea, D. 1974. Mapping at the age of three. *Journal of Geography* 73: 5–9.

Bloom, L. 1973. *One Word at a Time: The Use of Single Word Utterances before Syntax.* The Hague: Mouton.

Bloom, P., Peterson, M. A., Nadel, L., and Garrett, M. F., eds. 1996. *Language and Space.* Cambridge: MIT Press.

Bluestein, N., and Acredolo, L. 1979. Developmental changes in map-reading skills. *Child Development* 50: 691–97.

Bogartz, R. S., Shinskey, J. L., and Speaker, L. J. 1997. Interpreting infant looking: The event set * event set design. *Developmental Psychology* 33: 408–22.

Borke, H. 1975. Piaget's mountains revisited: Changes in the egocentric landscape. *Developmental Psychology* 11: 240–43.

Bowerman, M. 1989. Learning a semantic system: What role do cognitive predispositions play? In M. L. Rice and R. L. Schiefelbusch, eds., *The Teachability of Language.* Baltimore: Paul H. Brookes, pp. 133–69.

Bowerman, M. 1996. Learning how to structure space for language: A crosslinguistic perspective. In P. Bloom M. A. Peterson, L. Nadel., and M. F. Garrett, eds., *Language and Space.* Cambridge Ma.: MIT Press, pp. 385–436.

Bransford, J. D., Barclay, J. R., and Franks, J. J. 1972. Sentence memory: A constructive versus interpretive approach. *Cognitive Psychology* 3: 193–209.

Bremner, G. J. 1978a. Egocentric versus allocentric spatial coding in nine-month-old infants: Factors influencing the choice of code. *Developmental Psychology* 14: 346–55.

Bremner, G. J. 1978b. Spatial errors made by infants: Inadequate spatial cues or evidence of egocentrism? *British Journal of Psychology* 69: 77–84.

Bremner, J. G., and Bryant, P. E. 1977. Place versus responses as the basis of spatial errors made by young infants. *Journal of Experimental Child Psychology* 23: 162–71.

Bremner, J., Knowles, L., and Andreasen, G. 1994. Processes underlying young children's spatial orientation during movement. *Journal of Experimental Child Psychology* 57: 355–76.

Briggs, R. 1973. Urban cognitive distance. In R. M. Downs and D. Stea, eds., *Image and environment.* London: Arnold, pp. 361–88.

Brown, R. 1958. How shall a thing be called? *Psychological Review* 65: 14–21.

Bryant, D. J., and Tversky, B. 1992. Assessing spatial frameworks with object and direction probes. *Bulletin of the Psychonomic Society* 30: 29–32.

Bud, M. 1997. Effects of perceptual subdivision on the spatial estimation strategies of young children. Honors thesis. Tufts University, Medford, Ma.

Burgund, E. D., and Marsolek, C. J. 1998. Viewpoint-dependent priming for familiar objects in the right cerebral hemisphere., Poster presented at the Fifth Annual Conference of the Cognitive Neuroscience Society, University of Minnesota, Minneapolis, MN.

Bushnell, E. W., McKenzie, B. E., Lawrence, D. A., and Connell, S. 1995. The spatial coding strategies of one-year-old infants in a locomotor search task. *Child Development,* 66: 937–58.

Byrne, R. W. 1979. Memory for urban geography. *Quarterly Journal of Experimental Psychology* 31: 147–54.

Carey, S. 1991. Knowledge acquisition: Enrichment or conceptual change? In S. Carey and R. Gelman, eds., *The Epigenesis of Mind: Essays on Biology and Cognition.* Hillsdale NJ: Lawrence Erlbaum, pp. 257–91.

Carlson-Radvansky, L. A. 1995. Cooperation among spatial reference frames. Poster presented at the 36th Annual Meeting of the Psychonomic Society, Los Angeles, CA.

Carlson-Radvansky, L. A., and Jiang, Y. 1998. Inhibition accompanies reference-frame selection. *Psychological Science* 9: 386–91.

Carlson-Radvansky, L. A., and Radvansky, G. A. 1996. The influence of functional relations on spatial term selection. *Psychological Science* 7: 56–60.

Carpenter, P. A., Just, M. A., Keller, T. A., Eddy, W., and Thulborn, K. 1999. Graded functional activation in the visuospatial system with the amount of task demand. *Journal of Cognitive Neuroscience* 11: 9–24.

Carroll, M. 1993. Deictic and intrinsic orientation in spatial descriptions: A comparison between English and German. In J. Altarriba, ed., *Advances in Psychology, vol. 103. Cognition and Culture: A Cross-cultural Approach to Cognitive Psychology.* Amsterdam: North Holland/Elsevier Science, pp. 23–44.

Case, R., and Okamoto, Y. 1996. The role of central conceptual structures in the development of children's thought. *Monographs of the Society for Research in Child Development* 61: v-265.

Chabris, C. F., and Kosslyn, S. M. 1998. How do the cerebral hemispheres contribute to encoding spatial relations? *Current Directions in Psychological Science* 7: 8–14.

Chapman, M. 1988. *Constructive Evolution: Origins and Development of Piaget's Thought.* New York: Cambridge University Press.

Cheng, K. 1986. A purely geometric module in the rat's spatial representation. *Cognition* 23: 149–78.

Cheng, K., and Gallistel, C. R. 1984. Testing the geometric power of an animal's spatial representation. In H. L. Roitblatt, T. G. Bever, and H. S. Terrace, eds., *Animal Cognition.* Hillsdale NJ: Lawrence Erlbaum, pp. 409–23.

Choi, S., and Bowerman, M. 1991. Learning to express motion events in English and Korean: The influence of language-specific lexication patterns. *Cognition* 41: 83–121.

Choi, S., McDonough, L., Bowerman, M., and Mandler, J. 1999. Comprehension of spatial terms in English and Korean. *Cognitive Development* 14: 241–68.

Chomsky, N. 1965. *Aspects of the Theory of Syntax. Cambridge: MIT Press.*

Clearfield, M. W., and Mix, K. S. 1999. Number versus contour length in infants' discrimination of small visual sets. *Psychological Science* 10: 408–11.

Cornell, E. H., Heth, C. D., and Alberts, D. M. 1994. Place recognition and wayfinding by children and adults. *Memory and Cognition* 22: 633–43.

Cornell, E. H., Heth, C. D., and Broda, L. S. 1989. Children's wayfinding: Response to instructions to use environmental landmarks. *Developmental Psychology* 25: 755–64.

Cornell, E. H., Heth, C. D., and Rowat, W. L. 1992. Wayfinding by children and adults: Response to instructions to use look-back and retrace strategies. *Developmental Psychology* 28: 328–36.

Cousins, J. H., Siegel, A. W., and Maxwell, S. E. 1983. Wayfinding and cognitive mapping in large scale environments: A test of a developmental model. *Journal of Experimental Child Psychology* 35: 1–20.

Cramer, L., and Presson, C. 1995. Planning routes around obstacles: Does type of map information matter? Poster presented at the Biennial Meeting of the Society for Research in Child Development Indianapolis, IN.

Craton, L. G., Elicker, J., Plumert, J. M., and Pick, H. L., Jr. 1990. Children's use of frames of reference in communication of spatial location. *Child Development* 61: 1528–43.

Crawford, L. E., Regier, T., and Huttenlocher, J. (in press). Linguistic and non-linguistic organizations of space. *Cognition*.

Crowley, K., and Siegler, R. 1999. Explanation and generalization in young children's strategy learning. *Child Development* 70: 304–16.

Dalke, D. E. 1998. Charting the development of representational skills: When do children know that maps can lead and mislead? *Cognitive Development* 13: 53–72.

Deák, G. (in press). Hunting the fox of word learning: Why "constraints" fail to capture it. *Developmental Review*.

Dehaene, S. 1997. *The Number Sense: How the Mind Creates Mathematics.* New York: Oxford University Press.

DeLoache, J. S. 1984. Oh where, oh where: Memory-based searching by very young children. In C. Sophian, ed. *Origins of Cognitive Skills.* Hillsdale, N.J.: Lawrence Erlbaum, pp. 57–80.

DeLoache, J. S. 1987. Rapid change in the symbolic functioning of young children. *Science* 238: 1556–57.

DeLoache, J. S. 1989. Young children's understanding of the correspondence between a scale model and a larger space. *Cognitive Development* 4: 121–39.

DeLoache, J. S. 1991. Young children's understanding of models. In R. Fivush and J. Hudson, eds., *Knowing and Remembering in Young Children.* New York: Cambridge University Press, pp. 94–126.

DeLoache, J. S. 1995a. Early symbol understanding and use. *Psychology of Learning and Motivation* 33: 65–114.

DeLoache, J. S. 1995b. Early understanding and use of symbols: The model model. *Current Directions in Psychological Science* 4: 109–113.

DeLoache, J. S., and Brown, A. L. 1983. Very young children's memory for the location of objects in a large-scale environment. *Child Development* 54: 888–897.

DeLoache, J. S., and Burns, N. M. 1994. Early understanding of the representational function of pictures. *Cognition* 52: 83–110.

DeLoache, J. S., Kolstad, V., and Anderson, K. N. 1991. Physical similarity and young children's understanding of scale models. *Child Development* 62: 111–26.

DeLoache, J. S., Miller, K., Rosengren, K., and Bryant, N. 1997. The credible shrinking room: Very young children's performance with symbolic and nonsymbolic relations. *Psychological Science* 8: 308–13.

de Villiers, P. A. 1999. Language and thought: False complements and false beliefs. In J.G. de Villiers (chair) *Can Language Drive Conceptual Change?* Symposium conducted at the Biennial Meeting of the Society for Research on Child Development, Albuquerque, NM.

de Villiers, P. A., and de Villiers, J. G. 1974. On this, that, and the other: Non-egocentrism in very young children. *Journal of Experimental Child Psychology* 18: 438–447.

Diamond, A. 1985. Development of the ability to use recall to guide, as indicated by infants' performance on AB. *Child Development* 56: 868–83.

Diamond, A. 1990. Developmental time course in human infants and infant monkeys, and the neural bases of inhibitory control in reaching. In A. Diamond, ed., *The Development and Neural Bases of Higher Cognitive Functions.* New York: New York Academy of Sciences, pp. 637–76.

Diamond, A. 1991. Neuropsychological insights into the meaning of object concept development. In S. Carey and R. Gelman, eds., *The Jean Piaget Symposium Series: The Epigenesis of Mind: Essays on Biology and Cognition.* Hillsdale NJ: Lawrence Erlbaum, pp. 67–110.

Diamond, A. 1995. Evidence of robust recognition memory early in life even when assessed by reaching behavior. *Journal of Experimental Child Psychology* 59: 419–456.

Diamond, A., Cruttenden, L., and Neiderman, D. 1994. AB with multiple wells: I. Why are multiple wells sometimes easier than two wells? II. Memory or memory + inhibition? *Developmental Psychology* 30: 192–205.

Diamond, D. M., Fleshner, M., Ingersoll, N., and Rose, G. 1996. Psychological stress impairs spatial working memory: Relevance to electrophysiological studies of hippocampal function. *Behavioral Neuroscience* 110: 661–72.

Diedrich, F. J. 1997. Task dynamics of perseverative errors in infancy: Effects of load perturbations. Poster presented at the Meetings of the Society for Research in Child Development Washington, DC.

Diedrich, F. J., and Highlands, T. M. 1998. A dynamic theory of perseverative errors: Distinctive targets. Paper presented at the 11th Biennial International Conference on Infant Studies Atlanta, GA.

Diedrich, F. J., Thelen, E., and Smith, L. B. 1998. *Motor memory influences performance in the A-not-B task.* Paper presented at the 11th Biennial International Conference on Infant Studies, Atlanta, GA.

Diwadkar, V. A., and McNamara, T. P. 1997. Viewpoint dependence in scene recognition. *Psychological Science* 8: 302–07.

Dobbins, A. C., Jeo, R. M., Fiser, J., and Allman, J. M. 1998. Distance modulation of neural activity in the visual cortex. *Science* 281: 552–55.

Dow, G., and Pick, H. 1992. Young children's use of models and photographs as spatial representations. *Cognitive Development* 7: 351–63.

Downs, R. M. 1981. Maps and mappings as metaphors for spatial representation. In L. S. Liben, A. H. Patterson, and N. Newcombe, eds., *Spatial representation and behavior across the life span: Theory and application.* New York: Academic Press, pp. 143–66.

Downs, R. M. 1985. The representation of space: Its development in children and in cartography. In R. Cohen, ed., *The Development of Spatial Cognition.* Hillsdale NJ: Lawrence Erlbaum, pp. 323–45.

Downs, R. M., Liben, L. S., and Daggs, D. G. 1988. On education and geographers: The role of cognitive developmental theory in geographic education. *Annals of the Association of American Geographers* 78: 680–700.

Duva, C. A., Floresco, S. B., Wunderlich, G. R., Lao, T. L., Pinel, J. P. G., and Phillips, G. 1997. Disruption of spatial but not object-recogntion memory by neurotoxic lesions of the dorsal hippocampus in rats. *Behavioral Neuroscience* 11: 1184–96.

Edelman, G. M. 1992. *Bright Air, Brilliant Fire: On the Matter of the Mind.* New York: Basic Books.

Ellis, S., and Siegler, R. S. 1995. Developmental changes in children's understanding of principles and procedures of measurement. Paper presented at the Biennial Meeting of the Society for Research in Child Development, Indianapolis, IN.

Elman, J., Bates, E., Johnson, M., Karmiloff-Smith, A., Parisi, D., and Plunkett, K. 1996. *Rethinking Innateness: A Connectionist Perspective on Development.* Cambridge: MIT Press.

Etienne, A. S., Teroni, E., Maurer, R., Portenier, V., and Saucy, F. 1985. Short-distance homing in a small mammal: the role of exteroceptive cues and path integration. *Experientia* 41: 122–25.

Evans, G. W., Marrero, D. G., and Butler, P. A. 1981. Environmental learning and cognitive mapping. *Environment and Behavior* 13: 83–104.

Evans, G. W., and Pezdek, K. 1980. Cognitive mapping: Knowledge of real-world distance and location information. *Journal of Experimental Psychology: Human Learning and Memory* 6: 13–24.

Fabricius, W. V. 1988. The development of forward search planning in preschoolers. *Child Development* 59: 1473–1488.

Fabricius, W. V., and Wellman, H. M. 1993. Two roads diverged: Young children's ability to judge distance. *Child Development* 64: 399–414.

Farah, M. J. 1990. *Visual Agnosia: Disorders of Object-Recognition and What They Tell Us about Normal Vision.* Cambridge: MIT Press.

Farah, M. J., Hammond, K. M., Levine, D. N., and Calvanio, R. 1988. Visual and spatial mental imagery: Dissociable systems of representation. *Cognitive Psychology* 20: 439–62.

Farrell, M. J., and Robertson, I. H. 1998. Mental rotation and the automatic updating of body-centered spatial relationships. *Journal of Experimental Psychology: Learning, Memory, and Cognition* 24: 227–33.

Feigenson, L. and Spelke, E. 1998. Numerical knowledge in infancy: The number/mass distinction. Poster presented at the International Conference on Infant Studies Atlanta, GA.

Fillmore, C. J. 1975. *Santa Cruz Lectures on Deixis.* Bloomington: Indiana University Linguistics Club.

Fischer, K. W., and Bidell, T. 1991. Constraining nativist inferences about cognitive capacities. In S. Carey and R. Gelman, eds., *The Jean Piaget Symposium Series. The Epigenesis of Mind: Essays on Biology and Cognition.* Hillsdale, NJ: Lawrence Erlbaum, pp. 199–235.

Fisher, C. 1990. Children's discrimination of orientation: Development and determinants. In R. Vasta, ed., *Annals of Child Development.* Greenwich, CT: JAI, pp. 43–72.

Flavell, J. H. 1984. Discussion. In R. J. Sternberg, ed., *Mechanisms of Cognitive Development.* New York: W. H. Freeman, pp. 187–209.

Flavell, J. H., Omanson, R. C., and Latham, C. 1978. Solving spatial perspective taking problems by rule versus computation: a developmental study. *Developmental Psychology* 14: 462–73.

Fodor, J. A. 1983. *Modularity of Mind: An Essay on Faculty Psychology.* Cambridge: MIT Press.

Fodor, J. A. 1992. A theory of the child's theory of mind. *Cognition* 44: 283–96.

Forbus, K., Nielsen, P., and Faltings, B. 1991. Qualitative spatial reasoning: The CLOCK project. *Artificial Intelligence* 51: 1–3.

Ford, M. E. 1979. The construct validity of egocentrism. *Psychological Bulletin* 86: 1169–88.

Foreman, N., Arber, M., and Savage, J. 1984. Spatial memory in preschool infants. *Developmental Psychobiology* 17: 129–37.

Foster, T. C., Castro, C. A., and McNaughton, B. L. 1989. Spatial selectivity of rat hippocampal neurons: Dependence on preparedness for movement. *Science* 244: 1580–82.

Franklin, N., Henkel, L. A., and Zangas, T. 1995. Parsing surrounding space into regions. *Memory and Cognition* 23: 397–407.

Franklin, N., and Tversky, B. 1990. Searching imagined environments. *Journal of Experimental Psychology: General* 119: 63–76.

Franklin, N., Tversky, B., and Coon, V. 1992. Switching points of view in spatial mental models. *Memory and Cognition* 20: 507–18.

Friedman, N. P., and Miyake, A. (in press). Differential roles for visuospatial and verbal working memory in situation model construction. *Journal of Experimental Psychology: General.*

Frye, D., Zelazo, P. D., and Palfai, T. 1995. Theory of mind and rule-based reasoning. *Cognitive Development* 10: 483–527.

Fukusima, S. S., Loomis, J. M., and DaSilva, J. A. 1997. Visual perception of egocentric distance as assessed by triangulation. *Journal of Experimental Psychology: Human Perception and Performance* 23: 86–100.

Gallistel, C. R. 1990. *The Organization of Learning.* Cambridge: MIT Press.

Gauvain, M. 1993. The development of spatial thinking in everyday activity. *Developmental Review* 13: 92–121.

Gauvain, M. 1995. Thinking in niches: Sociocultural influences on cognitive development. *Human Development* 38: 25–45.

Gauvain, M., and Klaue, K. 1989. Influence of experience in an environment on the organization of directional information. Paper presented at the meetings of the Jean Piaget Society, Philadelphia, PA.

Gauvain, M., and Rogoff, B. 1986. Influence of the goal on children's exploration and memory of large-scale space. *Developmental Psychology, 22: 72–77.*

Gauvain, M., and Rogoff, B. 1989. Ways of speaking about space: The development of children's skill in communicating spatial knowledge. *Cognitive Development* 4: 295–307.

Geary, D. C. 1995. Reflections of evolution and culture in children's cognition: Implications for mathematical development and instruction. *American Psychologist* 50: 24–37.

Gelman, R., and Gallistel, C. R. 1978. *The Child's Understanding of Number.* Cambridge: Harvard University Press.

Gerstadt, C. L., Hong, Y. J., and Diamond, A. 1994. The relationship between cognition and action: Performance of children 3¤ to 7 years old on a Stroop-like day/night task. *Cognition* 53: 129–53.

Gilmore, R. O., and Johnson, M. H. 1995. Working memory in infancy: Six-month-olds' performance on two versions of the oculomotor delayed response task. *Journal of Experimental Child Psychology* 59: 397–418.

Gilmore, R. O., and Johnson, M. H. 1997. Egocentric action in early infancy: Spatial frames of reference for saccades. *Psychological Science* 8: 224–30.

Gladwin, T. 1970. *East Is a Big Bird.* Cambridge: Harvard University Press.

Glenberg, A. M., Meyer, M., and Lindem, K. 1987. Mental models contribute to foregrounding during text comprehension. *Journal of Memory and Language* 26: 69–83.

Glenn, M. J., and Mumby, D. G. 1998. Place memory is intact in rats with perirhinalcortex lesions. *Behavioral Neuroscience* 12: 1353–65.

Gnadt, J. W., and Andersen, R. A. 1992. Memory related motor planning activity in posterior parietal cortex of macaque. In S. M. Kosslyn and R. A. Andersen, eds., *Frontiers in Cognitive Neuroscience.* Cambridge: MIT Press, pp. 468–72.

Golbeck, S. L., Rand, M., and Soundy, C. 1986. Constructing a model of a large-scale space with the space in view: Effects on preschoolers of guidance and cognitive restructuring. *Merrill-Palmer Quarterly* 32: 187–203.

Goldin-Meadow, S., and Mylander, C. 1984. Gestural communication in deaf children: The effects and non-effects of parental input on early language development. *Monographs of the Society for Research in Child Development*, vol. 49 (3–4, serial 207).

Golinkoff, R. M., Hirsh-Pasek, K., and Hollich, G. (1999). Emergent cues for early word learning. In B. MacWhinney, ed., *The emergence of language.* Mahwah, NJ: Lawrence Erlbaum, pp. 305–29.

Goodale, M. A., and Milner, D. A. 1992. Separate visual pathways for perception and action. *Trends in Neuroscience* 15: 20–25.

Goodridge, J. P., and Taube, J. S. 1995. Preferential use of the landmark navigational system by head direction cells in rats. *Behavioral Neuroscience* 109: 49–61.

Gopnik, A., and Meltzoff, A. 1986. Words, plans, things, and locations: Interaction between semantic and cognitive development in the one-word stage. In S. A. Kuczaj II and M. D. Barrett, eds., *The Development of Word Meaning.* Berlin: Springer-Verlag.

Gopnik, A., and Meltzoff, A. 1987. The development of categorization in the second year and its relation to other cognitive and linguistic developments. *Child Development* 58: 1523–31.

Gopnik, A., and Meltzoff, A. N. 1997. *Words, Thoughts, and Theories*. Cambridge: MIT Press.

Gopnik, A., and Wellman, H.M. 1992. Why the child's theory of mind really is a theory. *Mind and Language* 7: 145–71.

Gopnik, A., and Wellman, H.M. 1994. The theory-theory. In L.A. Hirschfeld and S. Gelman, eds., *Mapping the Mind: Domain Specificity in Cognition and Culture*. New York: Cambridge University Press, pp. 257–93.

Graziano, M. S. A., Hu, X. T., and Gross, C. G. 1997. Coding the locations of objects in the dark. *Science* 277: 239–41.

Greenough, W., Black, J., and Wallace, C. 1988. Experience and brain development. *Child Development* 58: 539–59.

Haake, R. J., and Somerville, S. C. 1985. Development of logical search skills in infancy. *Developmental Psychology* 21: 176–86.

Haake, R. J., Somerville, S. C., and Wellman, H. M. 1980. Logical ability of young children in searching a large-scale environment. *Child Development* 51: 1299–1302.

Haith, M. M., and Benson, J. B. 1998. Infant Cognition. In D. Kuhn and R. Siegler, eds., *Handbook of Child Psychology: Cognition, Perception, and Language* vol. 2. New York: Wiley and Sons, pp. 199–254.

Halpern, E., Corrigan, R., and Aviezer, O. 1983. In, on, and under: Examining the relationship between cognitive and language skills. *International Journal of Behavioral Development* 6: 153–66.

Harris, P. L. 1994. Thinking by children and scientists: False analogies and neglected similarities. In L. A. Hirschfeld and S. Gelman, eds., *Mapping the Mind: Domain Specificity in Cognition and Culture*. New York: Cambridge University Press, pp. 294–315.

Hasher, L., and Zacks, R. T. 1979. Automatic and effortful processes in memory. *Journal of Experimental Psychology: General* 108: 356–88.

Hauser, M., and Carey, S. 1998. Building a cognitive creature from a set of primitives: Evolutionary and developmental insights. In C. Allen and D. Cummins, eds. *The Evolution of Mind*. Oxford: Oxford University Press, pp. 51–106.

Hayward, W. G., and Tarr, M. J. 1995. Spatial language and spatial representation. *Cognition* 55: 39–84.

Hazen, N. L., Lockman, J. J., and Pick, H. L. 1978. The development of children's representations of large-scale environments. *Child Development* 49: 623–36.

Herman, J. F., Shiraki, J. H., and Miller, B. S. 1985. Young children's ability to infer spatial relationships: Evidence from large familiar environments. *Child Development* 56: 1195–1203.

Herman, J. F., and Siegel, A. W. 1978. The development of cognition mapping of the large-scale environment. *Journal of Experimental Child Psychology* 26: 389–406.

Hermer, L., and Spelke, E. S. 1994. A geometric process for spatial reorientation in young children. *Nature* 370: 57–59.

Hermer, L., and Spelke, E. 1996. Modularity and development: a case of spatial reorientation. *Cognition* 61: 195–232.

Hespos, S. J., and Rochat, P. 1997. Dynamic mental representation in infancy. Unpublished manuscript, Psychology Department, University of Illinois at Urbana-Champaign, Urbana, IL.

Hirsh-Pasek, K., and Golinkoff, R. M. 1996. *The Origins of Grammar: Evidence from Early Language Comprehension*. Cambridge: MIT Press.

Hirtle, S. C., and Jonides, J. 1985. Evidence of hierarchies in cognitive maps. *Memory and Cognition* 13: 208–17.

Hoffman, D. D. 1993. No perception without representation. *Behavioral and Brain Sciences* 16: 247.

Hofstadter, M., and Reznick, J. S. 1996. Response modality affects human infant delayed-response performance. *Child Development* 67: 646–58.

Hollich, G. J., Hirsh-Pasek, K., and Golinkoff, R. M. 1999. Breaking the language barrier: An emergentist coalition model for the origins of word learning. Manuscript submitted for publication.

Horobin, K., and Acredolo, L. 1986. The role of attentiveness, mobility, history, and separation of hiding sites on Stage IV search behavior. *Journal of Experimental Child Psychology* 41: 114–27.

Hughes, M., and Donaldson, M. 1979. The use of hiding games for studying the coordination of viewpoints. *Educational Review* 31: 133–140.

Hunt, E., and Agnoli, F. 1991. The Whorfian hypothesis: A cognitive psychology perspective. *Psychological Review* 98: 377–89.

Hutchins, E. 1995. *Cognition in the Wild*. Cambridge: MIT Press.

Huttenlocher, J. 1974. The origins of language comprehension. In R. L. Solso, ed., *Theories in Cognitive Psychology*. New York: Wiley, pp. 331–368.

Huttenlocher, J. 1994. The emergence of number. Paper presented at the annual meeting of the Psychonomic Society, St. Louis, MO.

Huttenlocher, P. R. 1979. Synaptic density in human frontal cortex-developmental changes and effects of aging. *Brain Research* 163: 195–205.

Huttenlocher, J., and Gao, F. 1999. Coding of continuous quantity by infants. Paper presented at the biennial meeting of the Society for Research in Child Development, Albuquerque, NM.

Huttenlocher, J., and Hedges, L. V. 1994. Combining graded categories: Membership and typicality. *Psychological Review* 101: 157–65.

Huttenlocher, J., Hedges, L. V., and Duncan, S. 1991. Categories and particulars: Prototype effects in estimating spatial location. *Psychological Review* 98: 352–76.

Huttenlocher, J., Hedges, L. V., and Vevea, J. (in press). Why do categories affect stimulus judgment? *Journal of Experimental Psychology: General.*

Huttenlocher, J., Hedges, L. V., and Prohaska, V. 1988. Hierarchical organization in ordered domain: Estimating the dates of events. *Psychological Review* 95: 471–84.

Huttenlocher, J., Levine, S. C., and Jordan, N. 1994. A mental model for early arithmetic. *Journal of Experimental Psychology: General* 123: 284–96.

Huttenlocher, J., Levine, S., and Vevea, J. 1998. Environmental effects on cognitive growth: Evidence from time period comparisons. *Child Development* 69: 1012–29.

Huttenlocher, J., and Newcombe, N. 1984. The child's representation of information about location. In C. Sophian, ed., *Origins of cognitive skills*. Hillsdale, NJ: Lawrence Erlbaum, pp. 81–111.

Huttenlocher, J., Newcombe, N., and Sandberg, E. H. 1994. The coding of spatial location in young children. *Cognitive Psychology* 27: 115–48.

Huttenlocher, J., Newcombe, N., and Vasilyeva, M. 1999. Spatial scaling in young children. *Psychological Science* 10, 393–98.

Huttenlocher, J., and Presson, C. C. 1973. Mental rotation and the perspective problem. *Cognitive Psychology* 4: 277–99.

Huttenlocher, J., and Presson, C. C. 1979. The coding and transformation of spatial information. *Cognitive Psychology* 11: 375–94.

Iverson, J. M., and Goldin-Meadow, S. 1997. What's communication got to do with it? Gesture in children blind from birth. *Developmental Psychology* 33: 453–67.

Jammer, M. 1954. *Concepts of Space: The History of Theories of Space in Physics.* Cambridge: Harvard University Press.

Johnston, J. R. 1988. Children's verbal representation of spatial location. In J. Stiles-Davis, M. Kritchevsky, and U. Bellugi, eds., *Spatial Cognition. Brain Bases and Development.* Hillsdale, NJ: Lawrence Erlbaum, pp. 195–206.

Jonides, J., Smith, E. E., Koeppe, R. A., Awh, E., Minoshima, S., and Mintun, M. A. 1993. Spatial working memory in humans as revealed by PET. *Nature* 363: 623–25.

Jordan, N. C., Huttenlocher, J., and Levine, S. C. 1992. Differential calculation abilities in young children from middle- and low-income families. *Developmental Psychology* 28: 644–53.

Kagan, J. 1998. *Three Seductive Ideas.* Cambridge: Harvard University Press.

Kail, R. 1988. Developmental functions for speeds of cognitive processes. *Journal of Experimental Child Psychology* 45: 339–64.

Karmiloff-Smith, A. 1992. *Beyond Modularity: A Developmental Perspective on Cognitive Science.* Cambridge: MIT Press.

Karmiloff-Smith, A. 1998. Development itself is the key to understanding developmental disorders. *Trends in Cognitive Science* 2, 389–99.

Kaufman, J., and Needham, A. (in press). Objective spatial coding by 6.5-month-old infants in a visual dishabituation task. *Developmental Science.*

Keil, F. 1981. Constraints on knowledge and cognitive development. *Psychological Review* 88: 197–227.

Kellman, P. J., Gleitman, H., and Spelke, E. S. 1987. Object and observer motion in the perception of objects by infants. *Journal of Experimental Psychology: Human Perception and Performance* 13: 586–93.

Kellman, P. J., and Spelke, E. S. 1983. Perception of partly occluded objects in infancy. *Cognitive Psychology* 15: 483–524.

Kermoian, R., and Campos, J. J. 1988. Locomotor experience: A facilitator of spatial cognitive development. *Child Development* 59: 908–17.

Kim, C. K., Kalynchuk, L. E., Kornecook, T. J., Mumby, D. G., Dadgar, N. A., and Pinel, J. P. J. 1997. Object-recognition, spatial learning, and memory in rats prenatally exposed to ethanol. *Behavioral Neuroscience* 11: 985–95.

Klahr, D., and Robinson, M. 1981. Formal assessment of problem-solving and planning processes in preschool children. *Cognitive Psychology* 13: 113–48.

Knierim, J. J., Kudrimoti, H. S., and McNaughton, B. L. 1995. Place cells, head direction cells, and the learning of landmark stability. *Journal of Neuroscience* 15: 1648–59.

Kohler, S., Kapur, S., Moscovitch, M., Winocur, G., and Houle, S. 1995. Dissociation of pathways for object and spatial vision: A PET study in humans. *Neuroreports* 6: 1865–68.

Kosslyn, S. M. 1987. Seeing and imaging in the cerebral hemispheres: A computational approach. *Psychological Review* 94: 148–75.

Kosslyn, S. M., Heldmeyer, K. H., and Locklear, E. P. 1977. Children's drawings as data about internal representations. *Journal of Experimental Child Psychology* 23: 191–211.

Kosslyn, S. M., Pick, H. L., and Fariello, G. R. 1974. Cognitive maps in children and men. *Child Development* 45: 707–16.

Kretschmann, H.-J., Kammradt, G., Krauthausen, I., Sauer, B., and Wingert, F. 1986. Growth of the hippocampal formation in man. *Bibliotheca Anatomica* 28: 27–52.

Kuipers, B. 1982. The "map in the head" metaphor. *Environment and Behavior* 14: 202–20.

Laeng, B., Peters, M., and McCabe, B. 1998. Memory for locations within regions: Spatial biases and visual hemifield differences. *Memory and Cognition* 26: 97–107.

Landau, B. 1986. Early map use as an unlearned ability. *Cognition* 22: 201–23.

Landau, B. 1996. Multiple geometric representations of objects in languages and language learners. In P. Bloom, M. A. Peterson, L. Nadel., and M. F. Garrett, eds., *Language and Space*. Cambridge: MIT Press, pp. 317–64.

Landau, B., Gleitman, H., and Spelke, E. 1981. Spatial knowledge and geometric representation in a child blind from birth. *Science* 213: 1275–78.

Landau, B., and Jackendoff, R. 1993. "What" and "where" in spatial language and spatial cognition. *Behavioral and Brain Sciences* 16: 217–38.

Landau, B., and Spelke, E. 1988. Geometric complexity and object search in infancy. *Developmental Psychology* 24: 512–21.

Landau, B., Spelke, E., and Gleitman, H. 1984. Spatial knowledge in a young blind child. *Cognition* 16: 225–60.

Landis, T., Cummings, J. L., Benson, F. D., and Palmer, P. E. 1986. Loss of topographic familiarity: An environmental agnosia. *Archives of Neurology* 43: 132–36.

Langston, W., Kramer, D. C., and Glenberg, A. M. 1998. The representation of space in mental models derived from text. *Memory and Cognition* 26: 247–62.

Lansdale, M. W. 1998. Modeling memory for absolute location. *Psychological Review* 105: 351–78.

Laurendeau, M., and Pinard, A. 1970. *The Development of the Concept of Space in the Child*. New York: International Universities Press.

Lee, K., Eskritt, M., Symons, L. A., and Muir, D. 1998. Children's use of triadic eye gaze information for "mind reading." *Developmental Psychology* 34: 525–39.

Lee, T. R. 1970. Perceived distance as a function of direction in the city. *Environment and Behavior* 2: 40–51.

Lempers, J. D., Flavell, E. R., and Flavell, J. H. 1977. The development in very young children of tacit knowledge concerning visual perception. *Genetic Psychology Monographs* 95: 3–53.

Lepecq, J.-C., and Lafaite, M. 1989. The early development of position constancy in a non-landmark environment. *British Journal of Developmental Psychology* 7: 289–306.

Leslie, A. M. 1994. Pretending and believing: Issues in the theory of ToMM. *Cognition* 50: 211–38.

Levine, S., and Carey, S. 1982. Up front: The acquisition of a concept and a word. *Journal of Child Languages* 9: 645–57.

Levine, M., Jankovic, I. N., and Palij, M. 1982. Principles of spatial problem solving. *Journal of Experimental Psychology: General* 111: 157–75.

Levine, S. C., Jordan, N. C., and Huttenlocher, J. 1992. Development of calculation abilities in young children. *Journal of Experimental Child Psychology* 53: 72–103.

Levine, D. N., Warach, J., and Farah, M. J. 1985. Two visual systems in mental imagery: Dissociation of "what" and "where" in imagery disorders due to bilateral posterior cerebral lesions. *Neurology* 35: 1010–18.

Levinson, S. C. 1996. Frames of reference and Molyneux's question: Crosslinguistic evidence. In P. Bloom, M. A. Peterson, L. Nadel, and M. F. Garrett, eds., *Language and Space*. Cambridge: MIT Press, pp. 109–69.

Liben, L. 1981. Spatial representation and behaviour: Multiple perspectives. In L. S. Liben A. H. Patterson, and N. Newcombe, eds., *Spatial Representation and Behavior across the Life Span*. New York: Academic Press, pp. 3–36.

Liben, L. S. 1988. Conceptual issues in the development of spatial cognition. In J. Stiles-Davis, M. Kritchevsky, and U. Bellugi, eds., *Spatial Cognition: Brain Bases and Development*. Hillsdale, NJ: Lawrence Erlbaum, pp. 167–94.

Liben, L. S. (1999). Developing an understanding of external spatial representations. In I. E. Siegel, ed., *Development of Mental Representation: Theories and Applications*. Mahwah, NJ: Lawrence Erlbaum, pp. 297–321.

Liben, L. S. (in press). Thinking through maps. In M. Gattis, ed., *Spatial Schemes in Abstract Thought*. Cambridge: MIT Press.

Liben, L. S., and Downs, R. M. 1986. *Children's Production and Comprehension of Maps: Increasing Graphic Literacy*. Report G-83-0025. Washington DC: National Institute of Education.

Liben, L. S., and Downs, R. M. 1989. Understanding maps as symbols: The development of map concepts in children. In H. W. Reese, ed., *Advances in Child Development and Behavior*, vol. 22. New York: Academic Press, pp. 145–201.

Liben, L. S., and Downs, R. M. 1993. Understanding person-space-map relations: cartographic and developmental perspectives. *Developmental Psychology* 29: 739–52.

Liben, L. S., and Downs, R. M. (in press). Geography for young children: Maps as tools for learning environments. In S.L. Golbeck, ed., *Psychological Perspectives on Early Childhood Education*. Mahwah, NJ: Lawrence Erlbaum.

Liben, L. S., Moore, M. L., and Golbeck, S. L. 1982. Preschoolers' knowledge of their classroom environment: Evidence from small-scale and life-size spatial tasks. *Child Development* 53: 1275–84.

Liben, L. S., and Yekel, C. A. 1996. Preschoolers' understanding of plan and oblique maps: The role of geometric and representational correspondence. *Child Development* 67: 2780–96.

Light, L. L., and Zelinski, E. M. 1983. Memory for spatial information in young and old adults. *Developmental Psychology* 19: 901–906.

Linde, C., and Labov, W. 1975. Spatial networks as a site for the study of language and thought. *Language* 51: 924–39.

Lockman, J. J., and Pick, H. L., Jr. 1984. Problems of scale in spatial development. In C. Sophian, ed., *Origins of Cognitive Skills*. Hillsdale, NJ: Lawrence Erlbaum, pp. 3–26.

Loftus, G. 1978. Comprehending compass directions. *Memory and Cognition* 6: 416–22.

Logan, G. D. 1998. What is learned during automatization? II. Obligatory encoding of spatial location. *Journal of Experimental Psychology: Human Perception and Performance* 24: 1720–36.

Loomis, J. M., Klatzky, R. L., Golledge, R. G., and Philbeck, J. W. 1998. Human navigation by path integration. In R. G. Golledge, ed., *Wayfinding: Cognitive Mapping and Spatial Behavior*. Baltimore: Johns Hopkins Press.

Maki, R. H., Maki, W. S., and Marsh, L. G. 1977. Processing locational and orientational information. *Memory and Cognition* 5: 602–12.

Maki, R. H., and Marek, M. N. 1997. Egocentric spatial framework effects from single and multiple points of view. *Memory and Cognition* 25: 677–90.

Mandler, J. M. 1983. Representation. In P. H. Mussen, J. H. Flavell, and E. M. Markman, eds., *Handbook of Child Psychology: Cognitive Development* vol. 3. New York: Wiley, pp. 420–94.

Mandler, J. M. 1996. Preverbal representation and language. In P. Bloom, M. A. Peterson, L. Nadel., and M. F. Garrett, eds., *Language and Space*. Cambridge: MIT Press, pp. 365–84.

Mandler, J. M. (in press). On theory and modeling. *Developmental Science*.

Mandler, J. M., and McDonough, L. 1995. Long-term recall of event sequences in infancy. *Journal of Experimental Child Psychology* 59: 457–74.

Mandler, J., Seegmiller, D., and Day, J. 1977. On the coding of spatial information. *Memory and Cognition* 5: 10–16.

Mangan, P. A., Franklin, A., Tignor, T., Bolling, L., and Nadel, L. 1994. Development of spatial memory abilities in young children. *Society for Neuroscience Abstracts* 20: 363.

Mani, K., and Johnson-Laird, P. N. 1982. The mental representation of spatial descriptions. *Memory and Cognition* 10: 181–87.

Marcovitch, S., and Zelazo, P. D. 1998. A new meta-analysis of the A-not-B error. Poster presented at International Conference on Infant Studies, Atlanta, GA.

Mareschal, D., Plunkett, K., and Harris, P. 1995. Developing object permanence: A connectionist model. *Proceedings of the Seventeenth Annual Conference of the Cognitive Science Society*. Hillsdale, NJ: Lawrence Erlbaum.

Marzolf, D. P., and DeLoache, J. S. 1997. Search tasks as measures of cognitive development. In N. Foreman and R. Gillett, eds., *Handbook of Spatial Research Paradigms and Methodologies: Spatial Cognition in the Child and Adult* vol. 1. Hove, East Sussex: Psychology Press, pp. 131–52.

Marzolf, D. P., DeLoache, J. S., and Kolstad, V. 1997. The role of relational similarity in young children's understanding of scale models. Unpublished manuscript, Louisiana State University, Baton-Rouge, LA.

Matsuda, F. 1994. Concepts about interrelations amond duration, distance, and speed in young children. *International Journal of Behavioral Development* 17: 553–76.

Matthews, A., Ellis, A. E., and Nelson, C. A. 1996. Development of preterm and full-term infant ability on AB, recall memory, transparent barrier detour, and means-end tasks. *Child Development* 67: 2658–76.

McDonough, L. 1999. Early declarative memory for location. *British Journal of Developmental Psychology* 17: 381–402.

McKenzie, B. E., Day, R. H., and Ihsen, E. 1984. Localization of events in space: Young infants are not always egocentric. *British Journal of Developmental Psychology* 2: 1–9.

McNamara, T. P. 1991. Memory's view of space. In G. H. Bower, ed., *Advances in Research and Theory: The Psychology of Learning and Motivation*, vol. 27. San Diego, CA: Academic Press, pp. 147–86.

McNamara, T. P., and Diwadkar, V. 1997. Symmetry and asymmetry in human spatial memory. *Cognitive Psychology* 34: 160–90.

McNaughton, B. L., Barnes, C. A., Gerrard, J. L., Gothard, K., Jung, M. W., Knierim, J. J., Kudrimoti, H., Qin, Y., Skaggs, W. E., Suster, M., and Weaver, K. L. 1996. Deciphering the hippocampal polyglot: The hippocampus as a path integration system. *Journal of Experimental Biology* 199: 173–85.

McNaughton, B. L., Chen, L. L., and Markus, E. J. 1991. "Dead reckoning," landmark learning, and the sense of direction: a neurophysiological and computational hypothesis. *Journal of Cognitive Neuroscience* 3: 190–202.

McNaughton, B. L., Leonard, B., and Chen, L. 1989. Cortical- hippocampal interactions and cognitive mapping: a hypothesis based on reintegration of the parietal and inferotemporal pathways for visual processing. *Psychobiology* 17: 230–35.

Meck, W. H., and Church, R. M. 1983. A mode control model of counting and timing processes. *Journal of Experimental Psychology: Animal Behavior Processes* 9: 320–34.

Meltzoff, A. N. 1988. Infant imitation and memory: Nine-month-olds in immediate and deferred tests. *Child Development* 59: 217–25.

Meltzoff, A. N. and Moore, M. K. 1999. A new foundation for cognitive development in infancy: The birth of the representational infant. In E. K. Scholnick, K. Nelson, S. A. Gelman and P. H. Miller, eds., *Conceptual Development: Piaget's Legacy*. Mahwah, NJ: Lawrence Erlbaum, pp. 53–78.

Millar, S. 1988. Models of sensory deprivation: The nature/nurture dichotomy and spatial representation in the blind. *International Journal of Behavioral Development* 11: 69–87.

Millar, S. 1994. *Understanding and Representing Space: Theory and Evidence from Studies with Blind and Sighted Children*. Oxford: Clarendon Press.

Miller, E. K., and Desimone, R. 1994. Parallel neuronal mechanisms for short-term memory. *Science* 263: 520–22.

Miller, G. A., and Johnson-Laird, P. N. 1976. *Language and Perception*. Cambridge: Harvard University Press.

Miller, K. F. 1984. Child as the measurer of all things: Measurement procedures and the development of quantitative concepts. In C. Sophian, ed., *Origins of Cognitive Skills*. Hillsdale, NJ: Lawrence Erlbaum, pp. 193–228.

Miller, K. F. 1989. Measurement as a tool for thought: The role of measuring procedures in children's understanding of quantitative invariance. *Developmental Psychology* 25: 589–600.

Miller, K. F., and Baillargeon, R. 1990. Length and distance: Do preschoolers think that occlusion brings things together? *Developmental Psychology* 26: 103–14.

Miller, K. F., and Stigler, J. W. 1987. Counting in Chinese: Cultural variation in a basic cognitive skill. *Cognitive Development* 2: 279–305.

Mix, K. S., Huttenlocher, J., and Levine, S. C. 1996. Do preschool children recognize auditory-visual numerical correspondences? *Child Development* 67: 1592–1608.

Mix, K. S., Levine, S. C., and Huttenlocher, J. 1997. Numerical abstraction in infants: Another look. *Developmental Psychology* 33: 423–28.

Mizumori, S. J. Y. 1994. Neural representations during spatial navigation. *Current Directions in Psychological Science* 3: 125–29.

Monmonier, M. S. 1989. *Maps in the News: The Development of American Journalistic Cartography*. Chicago: University of Chicago Press.

Moore, C. 1996. Evolution and the modularity of mindreading. *Cognitive Development* 11: 605–21.

Morrongiello, B. A., Timney, B., and Humphrey, G. K. 1995. Spatial knowledge in blind and sighted children. *Journal of Experimental Child Psychology* 59: 211–33.

Morrow, D. G., Bower, G. H., and Greenspan, S. L. 1989. Updating situation models during narrative comprehension. *Journal of Memory and Language* 28: 292–312.

Morrow, D. G., Greenspan, S. L., and Bower, G. H. 1987. Accessibility and situation models in narrative comprehension. *Journal of Memory and Language* 26: 165–87.

Morrow, D. G., Leirer, V. O., Alieri, P. A., and Fitzsimmons, C. 1992. Creating situation models from narratives: Influence of age and spatial ability. Paper presented at the Second Annual Meeting of the Society for Text and Discourse, San Diego, CA.

Moscovitch, M., Kapur, S., Kohler, S., and Houle, S. 1995. Distinct neural correlates of visual long-term memory for spatial location and object identity: A positron emission tomography study in humans. *Procedures of the National Academy of Sciences USA* 92: 3721–25.

Muller, R. U., and Kubie, J. L. 1987. The effects of changes in the environment on the spatial firing of hippocampal complex-spike cells. *Journal of Neuroscience* 77: 1951–68.

Müller, U., and Overton, W. F. 1998. How to grow a baby: A re-evaluation of image-schema and Piagetian action approaches to representation. *Human Development* 41: 71–111.

Munakata, Y. (in press-a). Infant perseveration and implications for object permanence theories: A PDP model of the AB task. *Developmental Science*.

Munakata, Y. (in press-b). Infant perseveration: Rethinking data, theory, and the role of modelling. *Developmental Science*.

Munakata, Y., McClelland, J. L., Johnson, M. H., and Siegler, R. S. 1997. Rethinking infant knowledge: Toward and adaptive process account of successes and failures in object permanence tasks. *Psychological Review* 104: 686–713.

Munnich, E., Landau, B., and Dosher, B. A. 1997. Universals of spatial representation in language and memory. Poster presented at the 38th Annual Meeting of the Psychonomic Society, Philadelphia, PA.

Nadel, L. 1990. Varieties of spatial cognition: Psychological considerations. In A. Diamond, ed., *The Development of Neural Basis of Higher Cognitive Functions*. New York: New York Academy of Sciences, pp. 613–36.

Natoli, S., Boehm, R., Kracht, J., Lanegran, D., Monk, J., and Morrill, R. 1984. *Guidelines for Geographic Education: Elementary and Secondary Schools*. Washington, DC: Association of American Geographers, and Indiana, Pa: National Council for Geographic Education.

Naveh-Benjamin, M. 1987. Coding of spatial location information: An automatic process? *Journal of Experimental Psychology: Learning, Memory, and Cognition* 13: 595–605.

Nelson, K. E. 1974. Infants' short-term progress toward one component of object permanence. *Merrill-Palmer Quarterly* 20: 3–8.

Nelson, K. 1985. *Making Sense: The Acquisition of Shared Meaning*. Orlando FL: Academic Press.

Newcombe, N. 1988. The paradox of proximity in early spatial representation. *British Journal of Developmental Psychology* 6: 376–78.

Newcombe, N. 1989. Development of spatial perspective taking. In H. W. Reese, ed., *Advances in Child Development and Behavior*, vol. 22. San Diego, CA: Academic Press, pp. 203–247.

Newcombe, N. 1997. New perspectives on spatial representation: What different tasks tell us about how people remember location. In N. Foreman and R. Gillett, eds., *Handbook of Spatial Research Paradigms and Methodologies: Spatial Cognition in the Child and Adult*, vol. 1. Hove, East Sussex: Psychology Press, pp.85–102.

Newcombe, N., and Huttenlocher, J. 1992. Children's early ability to solve perspective-taking problems. *Developmental Psychology* 28: 635–43.

Newcombe, N., Huttenlocher, J., Drummey, A. B., and Wiley, J. 1998. The development of spatial location coding: Place learning and dead reckoning in the second and third years. *Cognitive Development* 13: 185–200.

Newcombe, N., Huttenlocher, J., and Learmonth, A. (in press). Infants' coding of location in continuous space. *Infant Behavior and Development*.

Newcombe, N., Huttenlocher, J., Sandberg, E., Lie, E., and Johnson, S. 1999. What do misestimations and asymmetries in spatial judgment indicate about spatial representation? *Journal of Experimental Psychology: Learning, Memory, and Cognition* 25: 986–96.

Newcombe, N., and Liben, L. S. 1982. Barrier effects in the cognitive maps of children and adults. *Journal of Experimental Child Psychology, 34: 46–58*.

O'Keefe, J., and Nadel, L. 1978. *The Hippocampus as a Cognitive Map*. Oxford: Clarendon Press.

O'Keefe, J., Nadel, L., Keightley, S., and Kill, D. 1975. Fornix lesions selectively abolish place learning in the rat. *Experimental Neurology* 48: 152–66.

O'Keefe, J., and Speakman, A. 1987. Single unit activity in the rat hippocampus during a spatial memory task. *Experimental Brain Research* 68: 1–27.

Ondracek, P. J., and Allen, G. L. 1997. Children's acquisition of spatial knowledge from verbal descriptions. Unpublished manuscript, University of South Carolina, Columbia.

Overman, W. H. 1990. Performance on traditional matching to sample, non-matching to sample, and object discrimination tasks by 12- to 32-month-old children: A developmental progression. In A. Diamond, ed.. *The Development of Neural Basis of Higher Cognitive Functions*. New York: New York Academy of Sciences, pp. 365–93.

Overman, W. H., Pate, B. J., Moore, K., and Peuster, A. 1996. Ontogeny of place learning in children as measured in the Radial Arm Maze, Morris Search Task, and Open Field Task. *Behavioral Neuroscience* 110: 1205–28.

Overton, W. F. 1998. Developmental psychology: Philosophy, concepts, and methodology. In R. M. Lerner, ed., *Handbook of Child Psychology. Theoretical Models of Human Development*, vol. 1. New York: Wiley, pp. 107–88.

Paris, S. G., and Mahoney, G. J. 1974. Cognitive integration in children's memory for sentences and pictures. *Child Development* 45: 633–42.

Park, D. C., and James, C. Q. 1983. Effect of encoding instructions on children's spatial and color memory: Is there evidence for automaticity? *Child Development* 54: 61–68.

Pascalis, O., de Haan, M., Nelson, C. A., and de Schonen, S. 1998. Long-term recognition memory for faces assessed by visual paired comparison in 3- and 6-month-old infants. *Journal of Experimental Psychology: Learning, Memory, and Cognition* 24: 249–60.

Pederson, E. 1995. Language as context, language as means: Spatial cognition and habitual language use. *Cognitive Linguistics* 6: 33–62.

Perlmutter, M., Hazen, N., Mitchell, D. B., Grady, J. C., Cavanaugh, J. C., and Flook, J. P. 1981. Picture cues and exhaustive search facilitate very young children's memory for location. *Developmental Psychology* 17: 109–110.

Perner, J. 1991. *Understanding the Representational Mind*. Cambridge: MIT Press.

Pezdek, K. 1983. Memory for items and their spatial locations by young and elderly adults. *Developmental Psychology* 19: 895–900.

Philbeck, J. W., Loomis, J. M., and Beall, A. C. 1997. Visually perceived location is an invariant in the control of action. *Perception and Psychophysics* 59: 601–12.

Piaget, J. 1951. *Play, dreams, and imitation in childhood* (C. Gattegno and F. M. Hodgson, trans.). New York: Norton.

Piaget, J. 1952. *The Origins of Intelligence in Children*. New York: Norton.

Piaget, J. 1954 *The Construction of Reality in the Child*. New York: Basic Books.

Piaget, J. 1941/1965. *The Child's Conception of Number*. New York: Norton.

Piaget, J., and Inhelder, B., 1948/1967. *The Child's Conception of Space*. (F. J. Langdon and J. L. Lunzer, trans.. New York: Norton.

Piaget, J., Inhelder, B., and Szeminska, A. 1960. *The Child's Conception of Geometry*. London: Routledge and Kegan Paul.

Pick, H. L., Jr. 1993. Organization of spatial knowledge in children. In N. Eilan, R. A. McCarthy, and B. Brewer, eds., *Spatial Representation: Problems in Philosophy and Psychology*. Oxford: Blackwell, pp. 31–42.

Pick, H. L., Jr., and Rieser, J. J. 1982. Children's cognitive mapping. In M. Potegal, ed., *Spatial Abilities: Development and Physiological Foundations*. San Diego, CA: Academic Press, pp. 107–28.

Pick, H. L., Jr., and Rosengren, K. S. 1991. Perception and representation in the development of mobility. In R. R. Hoffman and D. S. Palermo, eds., *Cognition and the Symbolic Processes: Applied and Ecological Perspectives*. Hillsdale, NJ: Lawrence Erlbaum, pp. 433–54.

Pinker, S. 1984. *Language Learnability and Language Development*. Cambridge: Harvard University Press.

Pinker, S. 1994. *The Language Instinct*. New York: Harper Collins.

Plumert, J. M., Carswell, C., DeVet, K., and Ihrig, D. 1995. The content and organization of communication about object locations. *Journal of Memory and Language* 34: 477–98.

Plumert, J. M., Ewert, K., and Spear, S. J. 1995. The early development of children's communication about nested spatial relations. *Child Development* 66: 959–69.

Plumert, J. M., and Hund, A. 1999. Distortions in memory for location: What role do spatial prototypes play?. Unpublished manuscript, University of Iowa.

Plumert, J. M., Pick, H. L., Marks, K. A., Kintsch, A. S., and Wegesin, D. 1994. Locating objects and communicating about locations: Organizational differences in children's searching and direction-giving. *Developmental Psychology* 30: 443–53.

Plumert, J. M., and Strahan, D. 1997. Relations between task structure and developmental changes in children's use of spatial clustering strategies. *British Journal of Developmental Psychology* 15: 495–514.

Presson, C. C. 1980. Spatial egocentrism and the effect of an alternative frame of reference. *Journal of Experimental Child Psychology* 29: 391–402.

Presson, C. C. 1982. Strategies in spatial reasoning. *Journal of Experimental Psychology: Learning, Memory, and Cognition* 8: 243–51.

Presson, C. C. 1987. The development of spatial cognition: Secondary uses of spatial information. In N. Eisenberg, ed., *Contemporary Topics in Developmental Psychology*. New York: Wiley, pp. 77–112.

Presson, C. C., and Hazelrigg, M. D. 1984. Building spatial representations through primary and secondary learning. *Journal of Experimental Psychology: Learning, Memory, and Cognition* 10: 723–32.

Presson, C. C., and Montello, D. R. 1994. Updating after rotational and translational body movements: Coordinate structure of perspective space. *Perception* 23: 1447–55.

Pufall, P. B. 1975. Egocentrism in spatial thinking: It depends upon your point of view. *Developmental Psychology* 11: 297–303.

Pufall, P. B., and Shaw, R. E. 1973. Analysis of the development of children's spatial reference systems. *Cognitive Psychology* 5: 151–75.

Quinn, P. C. 1994. The categorization of above and below spatial relations by young infants. *Child Development* 65: 58–69.

Quinn, P. C., Cummins, M., Kase, J., Martin, E., and Weissman, S. 1996. Development of categorical representations for above and below spatial relations in 3- to 7-month-old infants. *Developmental Psychology* 32: 942–50.

Radziszewska, B., and Rogoff, B. 1988. Influence of adult and peer collaborators on children's planning skills. *Developmental Psychology* 24: 840–48.

Radziszewska, B., and Rogoff, B. 1991. Children's guided participation in planning imaginary errands with skilled adult or peer partners. *Developmental Psychology* 27: 381–89.

Ratcliff, G. 1982. Disturbances of spatial orientation associated with cerebral lesions. In M. Potegal, ed., *Spatial Abilities: Development and Physiological Foundations*. New York: Academic Press, pp. 301–31.

Rider, E. A., and Rieser, J. J. 1988. Pointing at objects in other rooms: Young children's sensitivity perspective after walking with and without vision. *Child Development* 59: 480–94.

Rieser, J. J. 1979. Spatial orientation in six-month-old infants. *Child Development* 50: 1078–87.

Rieser, J. J. 1989. Access to knowledge of spatial structure at novel points of observation. *Journal of Experimental Psychology: Learning, Memory, and Cognition* 15: 1157–65.

Rieser, J., Doxsey, P., McCarrell, N., and Brooks, P. 1982. Wayfinding and toddlers' use of information from an aerial view. *Developmental Psychology* 18: 714–20.

Rieser, J. J., and Frymire, M. 1995. Locomotion with vision is coupled with knowledge of real and imagined surroundings. Paper presented at the 36th Annual Meeting of the Psychonomic Society, Los Angeles, CA.

Rieser, J. J., Guth, D. A., and Hill, E. W. 1986. Sensitivity to perspective structure while walking without vision. *Perception, 15:* 173–88.

Rieser, J. J., and Heiman, M. L. 1982. Spatial self-reference systems and shortest-route behavior in toddlers. *Child Development* 53: 524–33.

Rieser, J. J., Hill, E. W., Talor, C. R., Bradfield, A., and Rosen, S. 1992. Visual experience, visual field size, and the development of nonvisual sensitivity to the spatial structure of outdoor neighborhoods explored by walking. *Journal of Experimental Psychology: General* 121: 210–21.

Rieser, J. J., and Rider, E. A. 1991. Young children's spatial orientation with respect to multiple targets when walking without vision. *Developmental Psychology* 27: 97–107.

Rinck, M., and Bower, G. H. 1995. Anaphora resolution and the focus of attention in situation models. *Journal of Memory and Language* 34: 110–31.

Rinck, M., Hahnel, A., Bower, G. H., and Glowalla, U. 1997. The metrics of spatial situation models. *Journal of Experimental Psychology: Learning, Memory, and Cognition* 23: 622–37.

Rizzolatti, G., Fadiga, L., Fogassi, L., and Gallese, V. 1997. The space around us. *Science* 277: 190–91.

Roberts, R. J., and Aman, C. J. 1993. Developmental differences in giving directions: Spatial frames of reference and mental rotation. *Child Development* 64: 1258–1270.

Rochat, P., and Hespos, S. J. 1996. Tracking and anticipation of invisible spatial transformations by 4- to 8-month-old infants. *Cognitive Development* 11: 3–17.

Rogoff, B. 1990. *Apprenticeship in Thinking: Cognitive Development in Social Context.* New York: Oxford University Press.

Rogoff, B., and Lave, J., eds. 1984. *Everyday Cognition: Its Development in Social Context.* Cambridge: Harvard University Press.

Roskos-Ewoldsen, B., McNamara, T. P., Shelton, A. L., and Carr, W. S. 1998. Mental representations of large and small spatial layouts are orientation dependent. *Journal of Experimental Psychology: Learning, Memory and Cognition* 24: 215–26.

Rosser, R. A. 1983. The emergence of spatial perspective taking: An information-processing alternative to egocentrism. *Child Development* 54: 660–68.

Rudy, J. W., Stadler-Morris, S., and Albert, P. 1987. Ontogeny of spatial navigation behaviors in the rat: Dissociation of proximal- and distal-cue-based behaviors. *Behavioral Neuroscience* 101: 62–73.

Russell, J. 1995. At two with nature: Agency and the development of self-world dualism. In J. L. Bermudez, A. Marcel, and N. Eilan, eds., *The Body and the Self.* Cambridge: MIT Press, pp. 127–51.

Sadalla, E. K., Burroughs, W. J., and Staplin, L. J. 1980. Reference points in spatial cognition. *Journal of Experimental Psychology: Human Learning and Memory* 6: 516–28.

Sandberg, E. H. 1995. Development of spatial representation: The emergence of hierarchical organization. Doctoral dissertation Department of Psychology, University of Chicago, Chicago.

Sandberg, E. H., and Huttenlocher, J. 1997. Advanced spatial skills and advance planning: Components of 6-year-olds' navigational map use., Presented at the Biennial Meeting of the Society for Research in Child Development Washington, DC.

Sandberg, E. H., Huttenlocher, J., and Newcombe, N. 1996. The development of hierarchical representation of two-dimensional space. *Child Development,* 67: 721–39.

Save, E., and Moghaddam, M. 1996. Effects of lesions of the associative parietal cortex on the acquisition and use of spatial memory in egocentric and allocentric navigation tasks in the rat. *Behavioral Neuroscience* 110: 74–85.

Saxe, G. B., Guberman, S. R., and Gearhart, M. 1987. Social processes in early number development. *Monographs of the Society for Research in Child Development,* vol. 52 (2, Serial 216.

Schacter, D. L., and Nadel, L. 1991. Varieties of spatial memory: A problem for cognitive neuroscience. In R. G. Lister and H. J. Weingartner, eds., *Perspectives on Cognitive Neuroscience*. New York: Oxford University Press, pp. 165–85.

Schenk, F. 1985. Development of place navigation in rats from weaning to puberty. *Behavioral and Neural Biology* 43: 69–85.

Schiff, W. 1983. Conservation of length redux: A perceptual-linguistic phenomenon. *Child Development* 54: 1497–1506.

Schober, M. F. 1993. Spatial perspective-taking in conversation. *Cognition* 47: 1–24.

Schober, M. F. 1995. Speakers, addressees, and frames of reference: Whose effort is minimized in conversations about locations. *Discourse Processes* 20: 219–47.

Schober, M. F., and Bloom, J. E. 1995. The relative ease of producing egocentric, addressee-centered, and object-centered spatial descriptions. Poster presented at the 36th Annual Meeting of the Psychonomic Society, Los Angeles, CA.

Scholnick, E. K., Fein, G. G., and Campbell, P. F. 1990. Changing predictors of map use in wayfinding. *Developmental Psychology* 26: 188–93.

Schouela, D. A., Steinberg, L. M., Leveton, L. B., and Wapner, S. 1980. Development of the cognitive organization of an environment. *Canadian Journal of Behavioral Science* 12: 1–16.

Schuberth, R. E., Werner, J. S., and Lipsitt, L. P. 1978. The stage IV error in Piaget's theory of object concept development: A reconsideration of the spatial localization hypothesis. *Child Development* 49: 744–48.

Seress, L. 1992. Morphological variability and developmental aspects of monkey and human granule cells: Differences between the rodent and primate dentate gyrus. In C. E. Ribak, C. M. Gall, and I. Mody, eds., *The Dentate Gyrus and its Role in Seizures*. Amsterdam: Elsevier Science, pp. 3–28.

Shanon, B. 1984. Room descriptions. *Discourse Processes* 7: 225–55.

Shelton, A. L., and McNamara, T. P. 1997. Multiple views of spatial memory. *Psychonomic Bulletin and Review:* 4 102–106.

Shelton, A. L., and McNamara, T. P. 1999. *Systems of spatial reference in human memory.* Manuscript submitted for publication.

Shepard, R. N., and Cooper, L. A. 1982. *Mental Images and Their Transformations*. Cambridge: MIT Press.

Sholl, M. J. 1987. Cognitive maps as orienting schemata. *Journal of Experimental Psychology: Learning, Memory, and Cognition* 13: 615–28.

Sholl, M. J. 1995. The representation and retrieval of map and environment knowledge. *Geographical Systems* 2: 177–95.

Sholl, M. J., and Friedman, O. 1998. Place learning: Factors contributing to orientation-free performance. Presented at Psychonomics Society, Philadelphia, PA.

Sholl, M. J., and Nolan, T. L. 1997. Orientation specificity in representations of place. *Journal of Experimental Psychology: Learning, Memory, and Cognition* 23: 1494–1507.

Siegel, A. W., Herman, J. F., Allen, G. L., and Kirasic, K. C. 1979. The development of cognitive maps of large- and small- scale spaces. *Child Development* 50: 582–585.

Siegel, A. W., and Schadler, M. 1977. The development of young children's spatial representations of their classrooms. *Child Development* 48: 388–94.

Siegel, A. W., and White, S. H. 1975. The development of spatial representations of large-scale environments. In H. W. Reese, ed., *Advances in Child Development and Behavior*, vol. 10. New York: Academic Press, pp. 9–55.

Siegler, R. S. 1996. *Emerging Minds: The Process of Change in Children's Thinking*. New York: Oxford University Press.

Siegler, R. S., and Richards, D. D. 1979. Development of time, speed, and distance concepts. *Developmental Psychology* 15: 288–98.

Siegler, R. S., and Shipley, C. 1995. Variation, selection, and cognitive change. In T. Simon and G. Halford, eds., *Developing Cognitive Competence: New Approaches to Process Modeling*. Hillsdale, NJ: Lawrence Erlbaum, pp. 31–76.

Simon, T. J. 1997. Reconceptualizing the origins of number knowledge: A "non-numerical" account. *Cognitive Development* 12: 349–72.

Simons, D. J., and Wang, R. F. 1998. Perceiving real-world viewpoint changes. *Psychological Science* 9: 315–20.

Slobin, D. I. 1993. Is spatial language a special case? *Behavioral and Brain Sciences* 16: 249–51.

Smith, L. B., McLin, D., Titzer, B., and Thelen, E. 1995. *The task dynamics of the A not-B error*. Presented at the Biennial Meeting of the Society for Research in Child Development, Indianapolis, IN.

Smith, L. B., and Scheier, C. (in press). Babies have bodies: Why Munakata's net fails to meet its own goals. *Developmental Science*.

Smith, L. B., Thelen, E., Titzer, R., and McLin, D. 1999. Knowing in the context of acting: The task dynamics of the A-not-B error. *Psychological Review* 106: 235–60.

Sonnenschein, S., and Whitehurst, G. J. 1984. Developing referential communication: A hierarchy of skills. *Child Development* 55: 1936–45.

Sophian, C. 1984. Spatial transpositions and the early development of search. *Developmental Psychology* 20: 21–28.

Sophian, C., and Yengo, L. 1985. Infants' search for visible objects: Implications for the interpretation of early search errors. *Journal of Experimental Child Psychology* 40: 260–78.

Spelke, E. S., and Newport, E. L. 1998. Nativism, empiricism, and the development of knowledge. In R. M. Lerner, ed., *Handbook of Child Psychology: Theoretical Models of Human Development*, vol. 1. New York: Wiley, pp. 275–340.

Spelke, E. P., Vishton, P. M., and von Hofsten, C. 1995. Object perception, object-directed action, and physical knowledge in infancy. In M. S. Gazzaniga, ed., *The Cognitive Neurosciences*. Cambridge: MIT Press, pp. 165–79.

Spencer, J., Smith, L. B., and Thelen, E. 1997. Tests of a dynamic systems account of the A-not-B error: Perseverative reaching by 2- to 3-year-olds. Unpublished manuscript.

Stark, M., Coslett, B. H., and Saffran, E. M. 1996. Impairment of an egocentric map of locations: Implications for perception and action. *Cognitive Neuropsychology* 13: 481–523.

Starkey, P. 1992. The early development of numerical reasoning. *Cognition* 43: 93–126.

Starkey, P., and Cooper, R. G., Jr. 1980. Perception of numbers by human infants. *Science* 210: 1033–35.

Starkey, P., Spelke, E. S., and Gelman, R. 1990. Numerical abstraction by human infants. *Cognition* 36: 97–127.

Stevens, A., and Coupe, P. 1978. Distortions in judged spatial relations. *Cognitive Psychology* 10: 422–37.

Stryker, M. P. 1999. Sensory maps on the move. *Science* 284: 925–26.

Stulac, S. N., and Vishton, P. M. 1995. Using reaching to assess six-month-olds' understanding of the connectedness of a partly occluded object. Presented at the Biennial Meeting of the Society for Research in Child Development, Indianapolis, IN.

Talmy, L. 1983. How language structures space. In H. L. Pick Jr. and L. P. Acredolo, eds., *Spatial Orientation: Theory, Research, and Application*. New York: Plenum Press, pp. 225–82.

Tarr, M. J. 1995. Rotating objects to recognize them: A case study on the role of viewpoint dependency in the recognition of three-dimensional objects. *Psychonomic Bulletin and Review* 2: 55–82.

Tarr, M. J., and Pinker, S. 1989. Mental rotation and orientation-dependence in shape recognition. *Cognitive Psychology* 21: 233–82.

Taylor, H. L., and Klein, A. G. 1994. Referential communication of spatial location by children. Poster presented at the 35th Annual Psychonomics Conference, St. Louis, MO.

Taylor, H. A., Naylor, S. J., and Chechile, N. A. 1999. Goal-specific influences on the representation of spatial perspective. *Memory and Cognition* 27: 309–19.

Taylor, H. A., and Tversky, B. 1992a. Descriptions and depictions of environments. *Memory and Cognition* 20: 483–96.

Taylor, H. A., and Tversky, B. 1992b. Spatial mental models derived from survey and route descriptions. *Journal of Memory and Language* 31: 261–92.

Taylor, H. A., and Tversky, B. 1996. Perspective in spatial descriptions. *Journal of Memory and Language* 35: 371–91.

Thelen, E. and Smith, L. 1994. *A Dynamic Systems Approach to the Development of Cognition and Action.* Cambridge: MIT Press.

Thinus-Blanc, C., and Gaunet, F. 1997. Representation of space in blind persons: Vision as a spatial sense? *Psychological Bulletin* 121: 20–42.

Thorndyke, P. W., and Hayes-Roth, B. 1982. Differences in spatial knowledge acquired from maps and navigation. *Cognitive Psychology* 14: 560–89.

Titzer, B., Thelen, E., and Smith, L. 1995. The developmental dynamics of understanding transparency. Presented at the Biennial Meeting of the Society for Research in Child Development, Indianapolis, IN.

Tomasello, M., and Farrar, J. 1986. Joint attention and early language. *Child Development* 57: 1454–63.

Tversky, B. 1981. Distortions in memory for maps. *Cognitive Psychology* 13: 407–33.

Tversky, B. 1999. Talking about space. *Contemporary Psychology: APA Review of Books* 44: 39–40.

Tversky, B., and Clark, H. H. 1993. Prepositions aren't places. *Behavioral and Brain Sciences* 16: 252–53.

Tyler, D., and McKenzie, B. E. 1990. Spatial updating and training effects in the first year of human infancy. *Journal of Experimental Child Psychology* 50: 445–61.

Ungerleider, L. G., and Mishkin, M. 1982. Two cortical visual systems. In D. J. Ingle, M. A. Goodale, and R. J. W. Mansfield, eds., *Analysis of Visual Behavior.* Cambridge: MIT Press.

Uttal, D. H. 1994. Preschoolers' and adults' scale translations and reconstruction of spatial information acquired from maps. *British Journal of Developmental Psychology* 12: 259–75.

Uttal, D. H. 1996. Angles and distances: Children's and adults' reconstruction and scaling of spatial configurations. *Child Development* 67: 2763–79.

Uttal, D. H. 1999. *Seeing the big picture: Map use and the development of spatial cognition.* Unpublished manuscript Northwestern University, Evanston, IL.

Uttal, D. H., and Wellman, H. W. 1989. Young children's representation of spatial information acquired from maps. *Developmental Psychology* 25: 128–38.

Uzgiris, I. C., and Hunt, J. M. 1975. *Assessment in Infancy: Ordinal Scales of Psychological Development.* Urbana Illinois: University of Illinois Press.

Van der Linden, M., and Seron, X. 1987. A case of dissociation in topographical disorders: The selective breakdown of vector-map representation. In P. Ellen and C. Thinus-

Blanc, eds., *Cognitive Processes and Spatial Orientation in Animal and Man.* Boston: Martinus Nijhoff.

Wallace, J. R., and Veek, A. L. 1995. Children's use of maps for direction and distance estimation. Poster presented at the Biennial Meeting of the Society for Research in Child Development, Indianapolis, IN.

Wang, R. F., Hermer, L., and Spelke, E. S. 1999. Mechanisms of reorientation and object localization by children: A comparison with rats. *Behavioral Neuroscience* 113: 475–85.

Wang, R. F., and Simons, D. J. 1997. Layout change detection is differentially affected by display rotations and observer movements. *Investigative Ophthalmology and Visual Science* 38: s1009.

Wapner, S., Kaplan, B., and Ciottone, R. 1981. Self-world relationships in critical environment transitions: Childhood and beyond. In L. S. Liben, A. H. Patterson, and N. Newcombe, eds., *Spatial Representation and Behavior across the Life Span: Theory and Application.* New York: Academic Press, pp. 251–82.

Ward, S. L., Newcombe, N., and Overton, W. F. 1987. Turn left at the church, or three miles north: A study of direction giving and sex differences. *Environment and Behavior* 18: 192–213.

Warren, D. H. 1994. *Blindness and Children: An Individual Differences Approach.* New York: Cambridge University Press.

Waterhouse, L., Fein, D., and Modahl, C. 1996. Neurofunctional mechanisms in autism. *Psychological Review* 103: 457–89.

Webb, P. A., and Abrahamson, A. A. 1976. Stages of egocentrism in children's use of "this" and "that": a different point of view. *Journal of Child Language* 3: 349–67.

Wellman, H. M., Cross, D., and Bartsch, K. 1987. Infant search and object permanence: A meta-analysis of the A-not-B error. *Monographs of the Society for Research in Child Development* 51: 1–51.

Wellman, H. M., Fabricius, W. V., and Sophian, C. 1985. The early development of planning. In H. M. Wellman, ed., *Children's Searching: The Development of Search Skill and Spatial Representation.* Hillsdale, NJ: Lawrence Erlbaum, pp. 123–49.

Wellman, H. M., and Gelman, S. A. 1992. Cognitive development: Foundational theories of core domains. *Annual Review of Psychology* 43: 337–75.

Wellman, H. W., Somerville, S. C., and Haake, R. J. 1979. Development of search procedures in real-life spatial environments. *Developmental Psychology* 15: 530–42.

Wertsch, J. V., McNamee, G. D., McLane, J. B., and Budwig, N. A. 1980. The adult-child dyad as a problem-solving system. *Child Development* 51: 1215–21.

Wilcox, T., Nadel, L., and Rosser, R. 1996. Location memory in healthy preterm and full-term infants. *Infant Behavior and Development* 19: 309–23.

Wilcox, T., Rosser, R., and Nadel, L. 1994. Representation of object location in 6.5-month-old infants. *Cognitive Development* 9: 193–209.

Wilford, J. N. 1981. *The Mapmakers.* New York: Vintage Books.

Wilkening, F. 1981. Integrating velocity, time and distance information: A developmental study. *Cognitive Psychology* 13: 231–47.

Wilkening, F. 1982. Children's knowledge about time, distance, and velocity interrelations. In W. J. Friedman, ed., *The Developmental Psychology of Time.* New York: Academic Press.

Willats, P. 1999. Development of means-end behavior in young infants: Pulling a support to retrieve a distant object. *Developmental Psychology* 35: 651–67.

Woodin, M. E., and Allport, A. 1998. Independent reference frames in human spatial memory: Body-centered and environment-centered coding in near and far space. *Memory and Cognition* 26: 1109–16.

Wynn, K. 1992. Addition and subtraction by human infants. *Nature* 27: 749–50.

Xu, F., and Carey, S. 1996. Infants' metaphysics: The case of numerical identity. *Cognitive Psychology* 30: 111–53.

Yirmiya, N., Erel, O., Shaked, M., and Solomonica-Levi, D. 1998. Meta-analyses comparing theory of mind abilities of individuals with autism, individuals with mental retardation, and normally developing individuals. *Psychological Bulletin* 124: 283–307.

Zheng, W., and Knudsen, E. I. 1999. Functional selection of adaptive auditory space map by $GABA_A$-mediated inhibition. *Science* 284: 962–65.

Index